SURVIVING
the
FIREHOUSE

A Rookies Guide to Surviving the Firehouse and Fire Department Life

Mauro Porcelli

Printed in the United States of America

First Printing, 2018

Print ISBN: 978-1-54394-360-3

eBook ISBN: 978-1-54394-361-0

Ordering Information and Speaking Appearances

For details, contact the publisher at survivingthefirehouse@gmail.com

(352) 362-0487

TABLE OF CONTENTS

Greater love hath no man than this,
that a man lay down his life for his friends.

John 15:13

ACKNOWLEDGEMENTS

There is no doubt this book could not have been written if it were not for the patience and support of my wife Claudia and three children, Rachel, Nicholas and Andrew. My wife and I have been married for 25 years while raising our three beautiful children. Our kids have successful careers of their own. Nick works as a firefighter for the City of Orlando Fire department; Andrew is currently in firefighter training with Marion County Fire/Rescue, and Rachel works in the automotive industry. Thank you to my incredible family. They mean the world to me. In a way, I wrote this book with them in mind.

To all the wonderful firefighters I have had the honor to work with for so many years--you know who you are—I thank them for their guidance and encouragement throughout my career.

To all the firemen who constantly gave me life advice and tips on how to handle the obstacles in our way, but whose own lives were disasters, I know they meant well. I thank them for their efforts.

To all the firemen who said my dinners sucked and were too expensive, yet went back for seconds, and then asked me for the recipe, thanks for reminding me that actions speak louder than words.

To the several firefighters who had their Klondike Bars missing, didn't know who it was, and wanted an investigation, I confess. It was me.

To Orlando Local 1365, we wouldn't have what we have if it weren't for all their countless hours of hard work and dedication. Thank You.

To all the lieutenants and chief officers that I had the honor to work with and that took care of me, they are the best. Thank You.

To my editor, Gwenda Ward, for editing out my grammatical disasters in the editing process and making me sound like an academic genius, I thank her for her time and interest in this project.

INTRODUCTION

My reason for writing this book is to fill a void--to provide a blueprint or a road map to help the rookie or seasoned firefighter succeed in this career. The camaraderie among firefighters and the life-long friendships formed have no parallel, except maybe the Armed Forces. Sure, other co-workers bond over the course of their time together, but firefighters spend considerably more time in a twenty-four -hour span than most other professions. Much like the military, they also face life and death situations. Firehouse bonds are lasting. On the flip side, unfortunately, personality conflicts do arise. Sometimes you make life-long enemies--guys you never want to work with or see again--but that's life in the firehouse.

The advice in this book pertains to all firehouses around the world, not just Florida and other states. There are thousands of professional firefighters and firehouses around the world with various resources, manpower, and Standard Operating Procedures. The one constant for all of them is human nature.

No matter where a firefighter works, firefighters are the same everywhere. They respond to calls and work together to assist citizens with emergencies. On those calls, they function as a single unit. Over time, they bond. When they aren't on a call, they spend a lot of casual time together. During the down time, personalities can clash and egos take over. This book will guide you through the conflicts—dinner table conversations, workload sharing, clean-up duties, everything you will need to succeed.

This book is not just for rookies or someone thinking about getting into the fire department, but also for current firefighters as well. I will share my mistakes and career-enhancing decisions. I can't promise that you will get promoted to Chief or rise in rank, but I can guarantee that you will avoid the pitfalls most new firefighters make. You will have a greater understanding of how the fire department works and what makes it tick. Everything you need, you learn on the job; that's the way it's always been.

However, you can easily ruin your career and reputation if you do not know firehouse expectations. Hopefully, this book will help you significantly reduce the learning curve, give you a head start, or finely tune what you already know. It will guide you through your entire career, from rookie to retirement. It will give you a heads up on what to expect and advise you on how to deal with personality conflicts or work-related challenges.

I recently retired after twenty-five years as a professional firefighter. As I reflected on the career I loved and the many individuals I met along the way, I was struck by one question: why do some firefighters make it and others don't? I thought about why and how I survived life in the firehouse and others didn't. I thought about those who helped me succeed. I noticed others failed miserably. Why do so many individuals not bring good attitudes to the workplace? Often in the firehouse, I would see the new guys make serious mistakes. They wanted to do a good job but didn't have a proper guide. The profession lost many potentially good firefighters because of the absence of the right kind of leadership and guidance.

So, why am I writing this book to help people I don't even know? The answer is simple. Rookies need help. In the firehouse, getting along with your peers is critical; it will make or break your career. When I first started in the fire department back in 1988, I was asked, "What will you contribute to the fire department?" "What will you do to make a difference?" "Will you make the fire department a better place when you leave, than when you started?" Most of the time, firefighters are judged not by the lives they have saved or fires they've extinguished, but how they work with others in their down time. But no one stresses the importance of the interactions between the men and women with whom they work.

Throughout your career, you will be asked to pass your knowledge on to new firefighters. This could be to tell them about a certain call you were on, or, more simply, how to stretch a hose line effectively into a burning building. Whatever you want to pass on, is up to you, but these are the moments that contribute to your legacy. Your impact on your fellow firefighters is subtle but important.

Don't feel bad or depressed if your mistakes get rehashed as dinner time entertainment. This is the norm. You need to just suck it up. This is where personality and human nature take over. Will you laugh it off, or pout in a corner like a cry baby? I was always told, when you're working, you may have five hundred co-workers but only five true friends you can count on. Once you retire, your true friends might be only a few. My motive for this book is primarily to help others, especially those with whom I did not work with or that new person that is a lost soul who hasn't figured it out yet. It's natural to think that, upon retirement, the station will honor you and your contribution with a bronze statue in front of Fire Station 1. In reality, you are often forgotten by the time you clean out your locker. You really weren't as important as you thought.

As you read this book, you must keep an open mind about what you want to accomplish. Some people are fine being indifferent. Most of the time this will get you by. Since you are reading this book, you obviously want to improve yourself, be successful, and achieve your dream. What's important to remember is, most firefighters don't get fired from their job for lack of job performance; they get fired because of a bad attitude--not dealing well with coworkers, officers or the everyday firehouse routine. They take things personally, develop a bad attitude, and too soon, are up for a termination hearing.

Seriously, it's that sudden. One minute you're on top of the world getting your dream job, just bought a $50,000 truck that you really can't afford and married your prom date after moving out of your parents' basement. Next thing you know, you're fighting for your job because you didn't navigate your way through the fire department jungle. No worries! I'm here. This is why I wrote this book, for firefighters like you, who need a course correction, or the new guy who doesn't have a clue.

Unfortunately, many did not survive to retirement. These guys were short timers, here today and gone tomorrow. They didn't have a proper guide. These people, and many other excellent individuals, either quit or got fired early in their careers because they had a difficult time understanding the firehouse culture, or the reality of firehouse politics. There is no doubt that if they had a guide like this, many of these great people would have had a better chance of success. They would have understood firehouse expectations, survived their probation, and continued on in their careers.

Another reason for this book, and probably the main reason, is to help the young rookie and give him, or her, a shot at success. Time and again, I would see rookies try to fit in, but quickly get a bad reputation for being "lazy", "incompetent", "stupid", or some other denigrating term. Most of the time, the bad rap was no fault of their own, but their crew's for not taking the rookie under their wing and teaching him or her the do's and don'ts of the firehouse. No one is born a firefighter. The skills needed to fully comprehend the interactions and dynamics around the firehouse on a day to day basis are complex, especially when you are new. Learning to work with a crew of four to twenty-four strong personalities at the firehouse is tricky business and demands foreknowledge and skill.

Sometimes the bad raps are deserved because the rookie did not listen or did not care. From my experience, most of the time, failure was not their fault. The firehouse tends to have a pack mentality where everybody will gang up on the weaker person, especially the new guy. This, nonetheless, sets the rookie up for failure. Once a rookie has a bad reputation in the firehouse, it is almost impossible to undo the damage. Their lack of motivation, laziness, or whatever else the guys around the firehouse want to tag on them, will spread throughout the department, and they can rarely re-establish a good reputation. I always thought this was wrong and never truly reflected the new guy's potential. I have always felt it was a lack of leadership from the firefighters, to the lieutenant or captain in charge of the station. Most firefighters and officers claim to be leaders, but, where a lot of them fail miserably is in the mentorship area. They fail to nurture the young rookies. Don't get me wrong. There are many great mentors, but there should be many more. Yes, they may tell the novice what the job is, but there is a huge difference between that and establishing a relationship with the rookie. Some Firefighters claim it is common sense for the new person on the job to be the last to sit down at the dinner table and the first to start the dishes. They claim it is common sense for rookies to show up to work one hour early and start their station duties before anybody else shows up. They claim it is common sense for the rookie not to sit in the recliner at night with the guys and watch a movie with them; instead, he or she should be studying standard operating procedures in the kitchen and rules and regulations of the fire department, not hanging with the guys.

I have worked with hundreds of rookies over the years, and I assure you that none of this is common sense. Most people do not know any of these expectations unless their mentor tells them. These traits and habits should be taught by the crew, regardless of their age or experience. Once something is told, and the rookie still does not do what is expected, then he or she is fair game and must live with the consequences. Most of those hired, though, want to do a great job and impress their crew.

I promise you this: If you follow these guidelines, especially during orientation and your first days at the station, then you will have a successful rookie year and a successful career. I won't promise that all of your dreams will come true, but your adjustment to the firefighter's life will be smoother, and you will gain the respect of your peers, especially during your rookie year.

Understand something, you will always have those one or two assholes you will need to deal with at the firehouse. Being a rookie, by no means, is easy and will have its ups and downs. You need to set yourself up for success. You are not bigger or more important than the department you work for. It's true. We hate to admit it, but you can't ignore reality. As time passes, the people in your firehouse will move to other houses or retire. Your screw ups and accomplishments will be talked about less and less as attrition takes its toll.

This book is written for firefighters by a proud veteran firefighter who worked the streets in some of the roughest neighborhoods and busiest stations. No day shift or administrative duty for me! I tell you what you need to know in order to prepare you for the job ahead. I will not mince words. I don't worry about political correctness. I speak to you like I'm at the firehouse because I want you to "Survive the Firehouse". You need to understand the reality of your work life, straight and direct, not the romanticized, Hollywood version. I sometimes refer to a firefighter as a fireman, not fire woman. This book is for everyone but for convenience I use "fireman" to refer to both male and female firefighters. I tell you exactly what you need to hear.

A few words of warning about the use of profanity sprinkled throughout this book. Unfortunately, profanity is a fact of firehouse life. Leaving it out would give you the wrong view of the firehouse. I want to be realistic and honest. If you don't curse, that's fine; many people don't, but know in advance what to expect. Don't expect the firehouse to change because of you. I want you to "Survive the Firehouse." That is why I am here. I want you to learn to be at home there. You don't need to curse while talking at the station in order to fit in. Don't pretend to be somebody you are not in order to feel welcomed. Just be yourself.

This book covers your entire career from start to finish. It advises you about surviving orientation and probation, seeking promotion, working in small and large fire departments, and saving for retirement. It ends with recognizing when it's time to retire. In short, it includes everything a person needs to "Survive the Firehouse."

CHAPTER 1
THE FIREHOUSE CULTURE

If anything is true of the fire department, every fireman has an opinion; theirs is right and yours is wrong. Their "expertise" ranges from political, to marital advice, to picking up a hot date at a bar. Firemen will argue and fight with you about anything and everything. It's what they do. They are convinced that they are right, and your facts will not prove them wrong. We even had several individuals, divorced several times, giving marital advice. Guys who went bankrupt dished out financial advice. The best of all, the overweight dudes would give fitness advice. These shouting bouts are annoying if they target you and funny if they target someone else.

A prime example of this know-it-all attitude occurred during the Casey Anthony trial in Florida. Ms. Anthony was accused of killing her young child by putting her to sleep while she partied. It was a trial that most firemen followed closely, mainly because Casey Anthony was an attractive, young woman.

Plus, some of the guys were familiar with her from the local bar scene. Remarkably, throughout the trial, every fireman became an expert on criminal law and investigations. Instantly, through the magic of television, these guys knew everything about Case Law and how to present evidence to a jury. Last time I checked, none of the firefighters had law degrees.

Not every firefighter would agree with my characterization of the firehouse banter. Some might advise you differently. My response, if we were around the dinner table would be, "If you don't agree with me, write your own damn book!" No matter what situation or story is debated at the table, the guys are experts, and the person telling the story is the idiot. Topic doesn't matter.

This banter has a prominent place on the road map. I would be doing you a disservice if I didn't give you a head's up on what to expect and provide tips on how to deal with it. Trust me, you will lose many fights at the dinner table, but never give in, never say "I'm sorry". Stick to your guns.

I worked with some of the best people around and some of the biggest assholes this career has to offer. Around the firehouse, testosterone and egos abound, and bullshit is handed out more frequently than participation certificates during medal day. They are part of a firefighter's life. Forewarned is forearmed!

A firefighter's life can be hard, but it's rewarding if you learn from your mistakes. On occasion, I did things the hard way; I'm hoping to save you from the same mistakes. You cannot fight or change the culture around the firehouse. It can be the best playground an adult can hang out in, or your worst nightmare, especially if you are soft and fragile. Major word of advice: Never, ever let the crew see that you are mad or that their rhetoric affects you. Do not show any emotion whatsoever. If you do, then you're done. You will open up a pandora's box of whip ass and they will be relentless. Just smile, keep your mouth shut and move on.

CHAPTER 2
WORD OF WARNING: ATTITUDE

The next advice is true for most jobs, but especially true for firefighters: Attitude is first and foremost. You will be judged, not by your great God-given skills, or intelligence, but by your attitude. Nothing is worse from a new firefighter than dealing with a bad attitude. Nothing will get you into trouble faster, in school or work, than a bad attitude, or the belief that your co-workers, the department, or the world owes you something. Nobody owes you squat! You came to the department seeking a job; they did not seek you. You knew the pay before you started. You knew how busy you would be. You knew the benefits and the dangers of the job before you applied. If the fire department owes you anything, it's the best training and equipment available. This is what they promised you when you got hired (Let your union fight to improve the benefits. More about that in a later chapter).

An employee with a bad attitude is a cancer or poison to the crew and department. Its damage spreads and creates an environment that is difficult to work in. This is especially true in smaller departments and single company houses. Any officer will tell you that 80 percent of the job deals with personnel issues, and the remaining 20 percent actually deals with emergency calls. Most employees are reprimanded and terminated, not for lack of job performance, but for a bad attitude. Don't be that person.

We all have good and bad days; that's human nature. An individual will usually be marked as having a bad attitude if it's habitual. This is the guy who comes to work every day with a chip on his shoulder. He never wants to train, regularly argues with his officer, drains everyone's energy, and, in general, is high maintenance. Unfortunately, these people are not few and far between; they are quite common in every firehouse.

My best advice to you is to avoid these people at all costs. Their venom is contagious, and they want victims to come to their side. Worst of all, if you are the new guy, you really don't know any better. It is very easy for these people to taint you and inflict their poison on to you. Stay out of trouble by avoiding the bad attitude—don't bring it to the firehouse and avoid it when it's in the room.

You might ask, how do I avoid it when someone around me has a bad attitude? My best advice, when you are new, you need to realize that you do not have an opinion-- none, zilch. Don't even attempt to give one. If someone asks you what you think on any subject-- the union, administration or anything-- your response, with a humble smile, needs to be, "I'm just happy to be here." That's it! You're off the hook and everybody will think you're easy going, someone who likes to keep his mouth shut. If you keep getting pressed on an issue, just respond, "Hey, I'm the new guy. I really don't know much about that subject." Once again, you're off the hook and won't get pinned as the new guy that knows everything. Just sit back, keep your mouth shut, and listen. Trust me, you will be amazed at the stupidity that spews from some people's mouths. Makes one wonder how they got through life.

Too often, I would listen to these people in utter disbelief. Silently, I would wonder to myself, "My god, this guy is so wrong, so misinformed, so stupid, that he actually and truly believes himself!" On one occasion, I couldn't take it anymore. I tried to hold back but no longer could. I had to say something, like " Hey, Pauley, you're a complete freaking idiot. You have no idea what you are talking about, none!" "Did you make this shit up, or are you just that freaking stupid?" Then it began. Major "F" bombs flew back and forth. World War III began. Pauley continued to run his mouth and wouldn't admit that he truly was an idiot. Meanwhile, some of the guys jumped on the band wagon with their stupid crap. Before we knew it, it was dinner time where the argument intensified. Pauley continued to make himself look more and more stupid, while Joey D and Ricky P came in a close second. These kinds of open warfare happen more than I'd like to admit, but it was fun and comical. Best advice about handling these situations came from my friend and mentor, Chief Wayne Futch (Ret) who said, "Son, there is no sense of arguing with an idiot in front of an audience because after a while, the people standing around won't know who the idiot is." Best advice he ever gave.

CHAPTER 3
HOW 9/11
CHANGED EVERYTHING.

Before that horrible day of September 11, 2001, the fire department was just a basic organization where its main focus was fire suppression and EMS. Some departments dabbled in Hazardous Material (Haz-Mat) and Technical Rescue. After 9/11, everything changed in a monumental way. The focus of fire departments across the country went from fire suppression to a very heavy presence in Technical Rescue and Hazardous Materials mitigation. The US Government made billions of dollars in grant money available to most all fire departments. This money was designated for training and equipment. Almost overnight, fire departments everywhere acquired the best technical rescue gear and Haz-Mat monitoring equipment money could buy. This included all the tools necessary for building collapses and hazardous materials identification and mitigation. This focus persists today and impacts how fire departments train, think, and allot resources to the crews on the front line. If you love technical rescue and all it has to offer, then the fire department may suit you well. I will discuss later how you can work on the special units.

Frank Bello of 43 Engine, right and his son Frankie of 88 Engine, after a big fire in The Bronx. Two of FDNY's Bravest and Best!

CHAPTER 4
FIRE SCHOOL PREPARATION AND CHALLENGES

Fire school is the place where all the fun begins. A course in Fire Standards begins your journey as a firefighter. It is the first of a number of requirements to get hired. Most departments around the country now require candidates to also become EMT's or Paramedics before they even apply for a job. It's your first test as a person. It shows how you work with others and handle criticism.

Fire school can bring out the worst in a student. The new challenge may bring out your best or worst self, but you have to maintain a positive attitude, work hard and stay focused. Nobody is going to be perfect, and everybody will get screamed at. This is a very important time that you must be prepared for before you begin school.

When I went through fire standards back in 1988, the training took six weeks. Today, the training requires approximately fifteen intensive weeks. Depending on the state, standards will be plus or minus a few weeks. In fifteen weeks, instructors cover a lot more information. Today's fire college graduates are much more well-rounded than when I went through. I'm not going to include specific course work. You can investigate the various fire college requirements and programs on your own.

Advice: Get your ducks in a row before fire school. Some fire schools have a high dropout rate because of home life issues, boyfriend/ girlfriend disasters, lack of fitness, or inadequate mental preparation. These are all serious issues but should be addressed or fixed before facing the challenge of fire school. Today's fire school is no joke. You must be prepared mentally and physically. Any baggage you have before you begin must be taken care of if you are to have any chance of success.

When we talk about baggage, we have to mention everything. This includes drugs, alcohol, and tobacco products. Most fire schools will give you a drug screening test before you are accepted. Nowadays, it's the norm, and it should be. If you are a regular or casual user of marijuana or think you can skate by and nobody will know, think again. Don't delude yourself into thinking they won't find out or won't care. Fire departments around the country are serious about this and have a zero- tolerance policy during your pre-hire phase.

As far as tobacco products are concerned, Florida has a law called the "Firefighter Heart and Lung Bill." In Florida, when you are hired, the agency assumes you have a clean bill of health and that you do not or have not used tobacco products in the last few years. While on the job, and in the unfortunate event you contract lung or heart disease, it is assumed that you got that condition while working in the capacity of a firefighter. This is why you must sign a legal affidavit stating that you have not used any tobacco products within a specified period of time.

What else do they look for in a candidate? We have already established that a good attitude contributes to success in the firehouse. What else assures success? Being in top physical shape must be a priority. If you can run one mile and a half in ten minutes and do fifty push-ups and sit-ups, you are way ahead of most people, but this is just the beginning. Most fire schools are difficult. They are designed that way. You are entering an occupation where you must maintain peak physical shape to do your job effectively. Thousands of people graduate every year from fire school. The commonality with all of them is they were prepared mentally and physically.

So, what can you do to get ready? We already discussed attitude and excess baggage. Once you have this squared away, your physical fitness needs to be a top priority. Upper body strength and good endurance are a must. What works best? I have seen many workout fads come and go at the firehouse, but these days, "Cross Fit" seems to be the go to, golden standard with many firefighters. Other than hiring a high dollar personal trainer, you would be best served to go the cross fit route. Some people are animals at the gym. They may have some crazy routine that works for them. If that works for you and you are comfortable doing it, then stick with what works, but if you haven't found a solid fitness plan, try Cross Fit.

When my son Nicholas and Andrew decided to be a fireman, they both knew about the rigors of the job and what they needed to do physically to prepare for the course. Nick was in top shape. He wrestled in high school, did Mixed Martial Arts for several years, and regularly worked out in the gym. Andrew played football and stayed very active in sports. Both Nick and Andrew set their minds on something and went for it. After successfully completing the training, they both told me that doing "Cross Fit" helped them the most in preparation and fitness for the physical demands of the training.

Getting ready for Fire College

Cross Fit is designed to work your body as a whole, pushing you to your maximum limits. The gains in strength and endurance are impressive from their program. The occupation of firefighter is physically demanding and tends to lend itself well to people with upper body strength. Some women and short people, unfortunately, have a difficult time with upper body strength, especially when it comes to carrying and raising a ladder or dragging a victim out of a burning building. If you fall into this category, men included, then I believe Cross Fit, or something similar, will greatly improve your stamina and strength.

Around the country more and more fire schools and fire departments require a difficult test called, Candidate Physical Ability Test (CPAT). The CPAT is a nationally recognized physical agility test that students must pass in order to get into some fire schools and fire departments around the country. It is a test that the IAFF and the International Fire Chiefs Association has supported. This Candidate Physical Ability Test (CPAT) consists of eight separate events. The CPAT is a sequence of events requiring you to progress along a predetermined path from event to event in a continuous manner. This test was developed to allow fire departments to obtain pools of trainable candidates who are physically able to perform essential job tasks at fire scenes. (Fire Service Joint Labor Management Wellness/ Fitness Initiative)

Building Up endurance before school begins

Before you begin Fire Standards, make sure you are in the best mental and physical shape of your life. You don't want to be halfway through the program only to flunk out because you were too tired to continue. I advised both my sons that if you are so exhausted and you don't think you can continue, chances are you are almost done. Think about this for a second. You know you are already in top shape, probably the best shape of your life. There is not much out there that will exhaust you in the first few minutes of an evolution where you may want to quit. Chances are if you are at a breaking point where you can't continue, then you are probably almost done with that drill. You just have to reach deep inside a little longer to endure and prevail.

Some schools add their unique stamp on firefighter training. One such program is the Florida State Fire College, located in Ocala, Florida. It is notorious for its infamous "Ladder Tour." Make no mistake, it sucks! It sucked when I did it back in 1988 and still sucks now. This ladder tour is designed to get students to work together while exhausted and carrying ladders around the campus. Trainees are placed in groups of four to six, depending on the size of the ladder. The tour takes you through the woods, into windows, up stairwells or wherever the instructors feel would challenge the trainees.

Grinding it out during Ladder Tour

This tour has had its fair share of student drop outs. When Nick was doing his tour, unfortunately, a teammate of his didn't make it. He dropped the ladder, said he couldn't continue, and was kicked out of class. The sad thing was, he had almost completed the task. Just another five minutes and he would have finished the evolution. Just remember, these words: If you are so exhausted and you don't think you can continue, chances are you're almost done. If you survive the ladder torture, you're on your way.

Kalin Graham getting some Tech Rescue training.

CHAPTER 5
GRADUATION: BEFORE THE INTERVIEW

Your preparation for that first job interview actually begins in fire school. You may not believe it, but you are actually interviewing for your fire department job your first day in fire standards. When I talked about attitude in previous chapters, I warned against bad attitudes, personal baggage, and sloppy work habits. If you haven't fixed these personal attributes and improved your public persona, then do that before applying for that first job.

Happy to be done with Fire College

Think about it for a second. Here in Florida, most of the instructors are also active firefighters for the surrounding departments where the schools are located. The Florida State fire College in Ocala has instructors that work for Marion County, City of Ocala and the City of Gainesville fire departments. If you decide to attend fire standards in Seminole County, the instructors are active firefighters from the City of Orlando, Orange County and Seminole County fire departments. There is a good chance you may see your former instructors on the interview board.

At the very least, the fire Chief will ask one of the instructors about you. I have seen it countless times over the years. What if you griped about how hot it was out on the training ground or skipped a class or two? Or you didn't study hard enough for your tests? Your instructors probably will remember. It sucks, but it is reality. This is how it goes down, the chief asks, "Hey, Jimmy, what do you think about this guy or girl?" Jimmy replies, "Don't waste your time, Chief; he's a lazy ass with a bad attitude." Just that quickly, you lost the job.

Clean up your social media account

Be cautious with social media postings. Most employers search prospect's social media accounts. They believe that social media accounts are the best way to see who you really are, how you act, and what you value. If your pictures and videos show you doing things that would appall your mom, then get rid of it. Pictures of those really cool tattoos on your groin, you hitting a beer bong with friends, or sexually explicit selfies of your cleavage you posted a year ago, all these have to go. Even if you think it's no big deal, trust me, it is. If you're not sure, then delete everything. A few years later, you will be glad you did.

Taking a break after the big burn

CHAPTER 6
CHOOSING THE RIGHT FIRE DEPARTMENT FIT

You already made the decision to become a firefighter. You graduated from fire college and spent all of your time and effort to get to this point. What's next? Your next important decision to make is, "Where do I want to work?" Spend some time thinking about this one. You must consider many factors to make a wise and suitable decision here: Big city or small town? Saving babies from burning buildings or kittens from trees? Gunshot victims or pedestrian intoxication? Additionally, where do you want to raise your family? In a major city or small community? The choice you make at this juncture will impact the rest of your life.

Here in my home state of Florida, once certified as a Firefighter/EMT, you are qualified to work at any fire department in the state of Florida. Florida has a huge pool of municipalities to choose from. Consider carefully. The difference between departments is pretty significant. Most departments in my state are excellent places to work and lend themselves well to any personality. Here is an example.

The crew hanging out after doing the "Firefighter Combat Challenge"

The City of Orlando, Jacksonville, Tampa, and Miami along with Orange County Fire/Rescue, Metro Dade and Broward Counties are located in major metropolitan areas. These cities tend to have traffic congestion, over-populated, and have high crime rates. With all of this, these fire departments are extremely busy. On the flip side, you have departments such as the City of Ocala, Leesburg, and Lake County Fire Departments. These are also excellent departments that lend themselves well to people that want to stay busy, but not too crazy busy and still live in a somewhat small, safe community.

You must take several factors into consideration when making your decision. Do you want to raise your family in a big city with all the hustle and bustle? This may seem like a no brainer for someone starting out who is young and single. But, if you are married or thinking about starting a family, you have to consider your children's schools, playgrounds and home values. When you are young, an active night life is very important, along with all the amenities of a big city. You don't mind running twenty plus calls a shift, every shift, and dealing with a big city lifestyle.

On the other hand, many firefighters are born and raised in small towns. The thought of moving to a high-density city may cause nightmares. Smaller communities generally provide a more secure and family-friendly neighborhood atmosphere with relatively quiet fire departments. A lot can be said for small town charm. Working for a department with just one or two stations has its appeal. We all have our dream departments, regardless of where that may be.

So, now, with certificate in hand, it's time for applications and job offers. Blanket the state with applications. You have to apply everywhere to increase your odds of getting interviews and to increase your odds of getting hired. You need experience and, more importantly, a paycheck. Start somewhere. Do not get the mindset that you are too good to work for fire department A, and that you will wait until your dream department opens up. This is a huge mistake. Nothing will prepare you better for your dream department than working for another.

We are not born firefighters. Fire colleges can only prepare you so much. Before you get your dream job, you need to make your biggest mistakes early on, preferably in a place you don't plan on staying for long. You need to learn how to deal with co-workers, officers and all sorts of figurative grenades that will be thrown at you. The experience you gain from running EMS calls to working fires is priceless. These lessons are important to learn early on. Once you get your dream department, experience will make your assimilation much easier.

I have seen many firefighters sit around and wait for their dream department to open up. If the stars lined up perfectly, this wouldn't be a bad idea, but here's the problem. Most of these great departments only test every two to three years and consist of many parts. What if you failed the cutoff by just one point? The City of Orlando has multiple parts to its hiring process. In order to get on the list, you must pass the written test and then be invited to take the physical agility exam. Once you pass these two parts, you are then placed on Orlando's eligibility list. After this is completed, the Fire Chief will pick only a few to continue with the process. This process includes two interviews, two polygraphs, a full comprehensive medical exam, and a one thousand question psychological exam. Any of these can knock you out of the process if you do not meet the requirements of the department.

So, if you decide not to test with multiple departments and are willing to risk your career being set back two to three years, consider the above warning. My advice is to take a job! Any Job! Who cares where it is located? Get your career started, get the experience, and stay positive. If you stay clean, stay focused, stay out of trouble and maintain the highest physical conditioning possible, then I promise, eventually, your dreams will come true.

CHAPTER 7
PROS AND CONS OF A SMALL FIRE DEPARTMENT

First, let's talk about small fire departments, and the pros and cons from each. Before we begin, let me emphasize that firefighters in smaller departments are no less of a firefighter then they're bigger counterparts. I can assure you, as someone who has worked in both large and small departments, fires do not burn hotter in big departments than smaller ones. What is different, is the population is much greater and the buildings are taller and usually bigger, but that's about all

Pros of a Small Department

Working for a smaller department has many advantages. When you work for one, your actions usually affect the department more directly. You can serve on many different committees and effect change easier than in a big one. You can build a very tight knit unit with others from other stations and shifts.

You aren't lost in the crowd and can easily make a positive name for yourself.

Smaller departments can usually outfit their firefighters with updated gear and equipment more regularly then a bigger one. The main reason for this, is it is much easier for a fire chief to budget twenty to fifty new sets of bunker gear, helmets and boots than to budget several thousand. I know what you must be thinking, "But that chief has a much bigger budget for the bigger departments" Yes, that is correct, but everything--including fire apparatus, maintenance, and other departmental needs--is multiplied significantly by the greatly increased numbers of members, which makes the pool of money smaller for each line item.

Another positive about a smaller department is that firefighters get a lot more experience doing many different jobs on calls. There is one drawback. Smaller departments tend to have less man power than their bigger sister departments running the same type of call. We have to compare apples to apples here. In a smaller department, if you respond to a vehicle accident with reports of entrapment, chances are you may only have three firefighters on that unit. Some departments, believe it or not, still run two. If you only have three firefighters, chances are good that you are driving the engine, extricating the patient with your tools, and then having to perform critical ALS patient care. Sure, your extra manpower may be responding, but their distance could be a long way out, and you still may only get a few more people. Yes, there are many smaller departments that may have four firefighters on the engines, but that's not the norm.

In my old department where I started my career at Marion County Fire Depart, most of the time it was just two of us on duty at any one time. Fortunately, this now has changed. Before going to Orlando, I was fortunate enough to have gained lots of experience at different job functions. We definitely were not busier than Orlando, but I had to do everything from drive the fire truck, stretch the hose line and make entry into the building with just my partner and me. Oh yeah, and that was just my first week on the job. There is nothing that will benefit you more than being thrown to the wolves like that. You have no choice!

It's your job, and you do the best you can with what you have. Other times, I was on several one hundred plus acre brush fires with backup almost twenty minutes away. At small fire departments, with limited manpower, this could happen to you. We also ran into the same situation when we're running EMS calls with just two people. It wasn't all bad, though. I wouldn't give up that experience for anything. The lessons we learned, with such limited resources, were priceless. It's hard for firefighters to appreciate a severe lack of manpower until it happens to them.

Over the years, I realized that nothing builds confidence faster than working a trauma code by yourself. You better know your abilities because there is nobody to fall back on. If after a few years or months you decide to move on to a bigger department the experience and knowledge you just gained is huge. Your self-confidence will be high and you can continue with your career knowing you did something special.

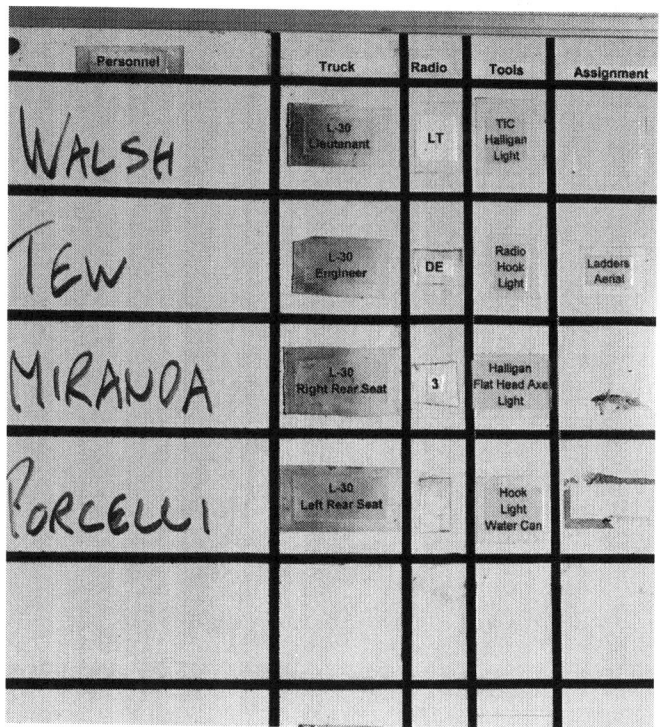

Personnel			Truck	Radio	Tools	Assignment
WALSH			L-30 Lieutenant	LT	TIC Halligan Light	
TEW			L-30 Engineer	DE	Radio Hook Light	Ladders Aerial
MIRANDA			L-30 Right Rear Seat	3	Halligan Flat Head Axe Light	
PORCELLI			L-30 Left Rear Seat		Hook Light Water Can	

Truck Company Riding Assignments

Cons of a Small Department

Now here's the downside of a small department. Keep in mind the pros and cons are subjective and would make for a nice debate, but I'm actually speaking from experience. The problem with smaller departments is that everybody knows your business. Your screw ups, regardless of how small, can easily grow into a huge mess. In the 80's and 90's, I lived in a modest sized community in Central Florida. The department was a 60-man department, relatively small and disorganized, with three shifts. If anyone did something wrong, there was no place to hide. If you pissed somebody off or were miserable at your firehouse, there were too few stations to transfer to. Even if you did find a spot, it was difficult to transfer, unless your officer wanted you gone.

Think about those implications? You had to have done something bad for your officer to want you gone. Whatever the facts, once your story reaches across town, it gets distorted and magnified. In a small town and small department, you might as well find a different career because it is sometimes near impossible to get your reputation back.

On another front, discipline at smaller departments can be random and inconsistent, at best. I don't mean discipline was not handed down, but it was heavy handed and inconsistent for the same offense. Administration would ignore union contract language, then threaten by memo, and pretty much do whatever they wanted to do with you. It was not uncommon for management to fire you first and ask questions later. Everything was "No Tolerance."

Don't get me wrong. What I'm explaining here was my experience with my old department. It was also common for me to hear the same stories from other firemen from other agencies across the state. I sincerely believe that the reason most smaller agencies battle this type of behavior is the fact that there is not as much collective strength in smaller departments than in larger ones.

Firefighters in smaller departments are a special breed, tough individuals, mentally and physically. They not only run a ton of calls with limited manpower, but they also deal with small town politics that can control the day to day decision making. They have to deal with it because there aren't as many firefighters to dilute the mandates coming from management. In larger departments, it's easy to say, "Who cares? That doesn't affect me directly because I am just one person out of one thousand." But in smaller departments, almost everything affects you, either directly or indirectly.

Here is another hazard regarding some smaller departments. There seems to be a revolving door with Fire Chiefs. I never quite understood that logic? Why would a City Manager or County Administrator hire someone that was fired from several other departments? This pattern seems to be quite common in some smaller departments around the country. Believe me, there are many larger ones which also deal with this issue, but it's more noticeable in the smaller ones. The large departments, for the most part, seem to promote from within the agency where people have been groomed for many years. Remember, there are many great Fire Chiefs and management teams at the smaller departments, but it seems the door revolves more around the smaller ones.

Something else to consider about working for a small fire department is the safety issue of limited manpower. Manpower is everything. It is arguably the most important factor that can make or break any department. When you have manpower, you have increased safety, efficiency and hands to protect you in case something goes wrong.

When I was in my old department, my number one concern was safety. I wondered what would happen if I was lost, trapped, or severely injured. Remember, it was just two or three of us on most fire scenes. Even the engineer was inside fighting fire. In addition, you had to do it quickly because there were no hydrants and only a limited water supply. If we were trapped under a roof or lost, we were screwed! Really, there was no help. If another station did respond, it was probably some unfit guy that showed up in flip flops and shorts from the neighboring volunteer station. Most of the time, you were lucky if he brought another engine or tanker. If you like a small town with a personal feel and don't want to get lost in the crowd, then a smaller department is the way to go.

CHAPTER 8
THE PROS AND CONS OF A BIG DEPARTMENT

For me personally, this is where it's at. Working for a big department was going to be the pinnacle of my career. There would be nothing like it. I could have a specific job, regardless of the type of call, was my dream come true. There would be no more free lancing on the scene or working with limited manpower. We would have real training, accountability, and best of all, tons of manpower. I would finally get to experience real department pride, where everyone was happy to be there.

Pros of Big City Departments

One positive aspect of the large city fire departments is that they don't have a manpower problem. It probably goes without saying, large metropolitan areas have bigger buildings and a greater variety of disasters that merit a call to the fire department. These skyscrapers are complicated structures and complicated to extricate victims in case of fire. The loss of life can be dramatic when disaster strikes.

In Orlando, we had massive apartment complexes that spanned entire city blocks with 100% occupancy rates. At night, these wooden structures, with common cock lofts (firefighter term for attic space with no fire breaks and therefore no barrier to contain the fire) and no fire sprinklers, would regularly catch fire. With hundreds of occupants, you better have manpower, proper training and the right people to take care of the job. With manpower, you can do almost anything.

This really comes into play when a department is designing Standard Operating Procedures, or SOP's. We in Orlando, as in most bigger departments, had excellent SOP's, one of the best in the country. Numerous departments outside the state modeled theirs after ours. Our SOP's have been revised and improved several times over the last hundred years. The final product is the envy of many large fire departments.

When talking about SOP's you have to compare apples to apples. Our SOP's will not work in Manhattan nor would Boston or Chicago's work for us. Building types and construction, and most important manpower, vary greatly and must be considered when designing your own SOP.

Orlando Pipes & Drums in Times Square

When I decided to work for Orlando or any other larger department that would hire me, one of the main things that I wanted was to have only one job function and not have twenty different roles. When I was assigned to "Tower 10 Right Jump Seat," my role, for any fire, was forceable entry and interior search. I knew I needed to come off the truck with my Irons and Pike Pole every single time. If it was a commercial building, I would grab the saw to cut the locks and gates to make sure the hose team would have easy access to the fire. If we were responding to a high rise, I knew I needed to bring the high-rise pack. This is what made my job simple; I knew exactly what I had to do for every single type of fire. Since Tower Trucks also did extrication work, I knew I needed to come off the truck with cribbing and the hydraulic cutters. For me, this is what I was looking for in a fire department—a chance to excel as a fireman.

Another pro for larger municipalities is that they pay their firefighters higher wages. We have to be careful with this to make sure we compare apples to apples. Larger cities tend to pay more money because the cost of living is higher. For example, South Florida has a much higher median home price and higher property taxes than North or Central Florida. Even though these larger municipalities pay more money, you have to compare the cost of living expenses.

Naples, Florida, is a prominent, high end, exclusive city with an excellent fire department. Its median home price is $320,000, (Naples Area Board of Realtors, 2018) with a very high property tax rate. The problem lies with the starting salary of just $45,755, (City of Naples Human Resources, 2015). It's difficult to live in the city while making just $46,000.

An ideal situation would be to live in a small town and commute to the larger city for work. That's what I did, and it worked for my family. This is actually very common and can be quite lucrative if done correctly. Where you need to be careful is to find out if your department has a distance rule to work. The City of Orlando Fire Department rule states that you cannot live more than forty-five minutes from the city line. Each fire department is different in how it enforces this rule, so due diligence would come in handy when making residence decisions.

When deciding where you want to work, financial considerations matter a great deal. Here is a simple actuary you can use. The chart uses an average small home price with an average person's monthly living expenses. It's definitely not scientific, but it will give you an idea about how to decide. This chart will show how much money you need to earn to survive.

BUDGET WITH $42,420 ANNUAL SALARY

$150,000 House@ 4.5% annual Interest 30 years

Mortgage	$760
Food	$600
Taxes	$125
Fun	$500
Insurance	$125
Electricity	$250
Savings	20% of Ck
Phone/Cable/ Internet	$125
Repairs	$100
Miscellaneous	$200
Total	$42,420 per/ yr.
	w/o savings

Vehicle $20,000

Payment	$400
Gas	$250
Insurance	$100

The Cons of a Big City Department

There are many positives to working for larger departments. But with that said, there are definitely a few disadvantages. One of the drawbacks is that it is difficult to bring about change. Large departments usually have been around for many years and have deeply rooted traditions embedded in them. To make changes is almost impossible. It takes a very long time to convince enough people that change is needed.

For example, Orlando had this thing called "rain birds". Try as I might, I couldn't get the large department to do away with them. These rain birds were large sprinkler heads that the Orlando fire department would have its firefighters lay out with thousands of feet of hose to saturate an area. These things were usually set up in an area where a brush fire just occurred. It didn't even have to be a big fire. Any fire could require us to set these things up. Every two hours we would go and check on the pump connected to the hydrant to make sure there was enough fuel. The worst part was, some poor guy had to pick up the thousands of feet of hose, mud encrusted and water logged.

When I was hired in Orlando, I didn't know much about high rises or large commercial building fires, but I did know brush fires better than anyone. In my home county we had massive brush fires we responded on a regular basis. I tried, to no avail, to convince the department that their Rain Bird strategy was ridiculous and needed to be scrapped. Their response was, "We have always done it this way." That was it; that was the very best answer I could get. After a few years, the old timers retired out, and then the rain birds eventually were taken out of service.

With these bigger departments, you really don't have much of a voice to effect positive change. That's not to say administration won't listen to you or that they don't like you, but because these departments are so large, change takes much longer and improvements can be delayed.

CHAPTER 9
PREPARING FOR YOUR DEPARTMENT TEST

Once you have decided where you want to work, you now need to prepare for your test. This test varies greatly from department to department. The City of Orlando Fire Department has an extensive hiring process with approximately 1000 candidates testing. The process usually begins with a written exam, followed by a fire ground practical, EMS practical, two polygraphs, psychological, two interviews and a very extensive medical exam. Most departments usually only give a written exam, practical, medical and one interview with the chief. Know the requirements for the department you are testing for and study up on how to best get through the process. Remember, any minor detail can knock you out of the running.

You should be preparing for the testing process years before you actually take it by making healthy life choices before you decided to be a firefighter. If you have any one of the following on your record, you will be knocked out of the running in any good department: Smoking, Drugs, Arrest Record, Bad Past Employer Recommendation, Past Terminations, DUI, Not Being a Good Person and Horrible Driving Record. My best advice is to lead a good clean life before deciding you want this career, and your chances at your dream department will rise substantially.

Your interview with the Chief

This is your opportunity to shine. It's your chance to convince the person who will hire you that you deserve that highly coveted position at their fire department. You only have one shot at this, so don't screw it up.

Some Dos and Don'ts

- Do give the Chief a good, strong hand shake. A weak hand is a weak person.
- Do sit up in your chair.
- Do speak clearly and look directly at the person asking you the questions.
- Do be confident, not cocky.
- Do be prompt for the interview. Yes, some people have been late.
- Don't be negative or talk about how your last department sucked

- Do dress up in your Sunday best. Suit and tie, nice dress but no cleavage.
- Do be clean-shaven and well groomed.
- Do not talk politics, gender issues or anything controversial.
- Don't ask about salary or time off. Do your homework before you apply.

The Interview

- Practice before your interview. Pick two or three people that you trust, preferably adults, and have them ask you the following questions:
- Why do you want to work for this fire department?
- Why should I hire you over somebody else?
- How will you contribute to this department and be a positive force?
- Tell me about your personal strengths and how that will help our department
- How do you get along with others?

Nick looking good before his big interview

Have your questioners come up with some spontaneous, off-the-wall questions. That way, you are ready and confident that you can answer anything. Don't answer with single word responses. Elaborate on your responses. Discuss the answer in a positive manner where it introduces a follow-up question. This starts a dialogue and lets the Chief know you are a people person who can communicate well with anyone.

If all goes well, you will pass the test, have a great interview and be offered a job. Soon you will begin to enjoy your life as a firefighter. Welcome to the firehouse! There's more to consider, but you are on your way.

CHAPTER 10
SHOULD I BE A PARAMEDIC?

Well, here's the deal. If you want to land your dream job, then you will significantly increase your chances of getting hired if you go to paramedic school. This is usually one year of extra schooling after you complete EMT. Today's fire department has changed; we like to think 99 percent of our calls are fire related where we are doing some pretty cool heroic stuff. The fact of the matter is, approximately 80 percent of a fire department's call load is EMS related. On the plus side, paramedics are highly utilized, and therefore, this training increases your likelihood of being hired. On the negative side, being a paramedic in a busy department is not easy.

Think about it, you are doing 80 percent of the work. You could be stuck on a busy rescue for years until you transfer or promote out. Paramedics have a high burnout rate. They tend to become disgruntled employees and develop sleeping issues on their days off after staying up for twenty-four hours straight.

But, like I stated earlier, it is very difficult to have your dream job without it.

Being a paramedic can be very rewarding. While working as one, you may actually save a life every shift. It's not as heroic as pulling a baby out of a burning home or rescuing a family from a burning building, but a street medic is truly where you can make a positive difference in another's life.

You actually can make some substantial money working as a Paramedic. On top of this, most departments have an additional stipend for being assigned to the rescue for the day.

Most departments will give you anywhere from $5,000-$10,000 a year in extra pay; plus, the stipend is usually pensionable. Paramedics are in demand when overtime becomes available. It's usually the medic spot that needs filling. Working for time and a half a few times per month is usually not a bad gig, considering the extra money you can make.

CHAPTER 11
ORIENTATION AND PROBATION

Congratulations! You made it through fire college, EMT or Paramedic school and the hiring process. You finally have your dream job. You've worked hard and deserve credit for your determination. Now that you're hired, life doesn't suddenly get simple. In fact, it gets complicated. Almost all fire departments provide an orientation—watch your step, this is where the land-mines hide. Remember, we are not born firefighters; we learn it. Don't rest on your achievement now. Your true test comes during your first year.

Finding your place in the firehouse and getting along with your co-workers may be your next challenge. What you need to remember is that almost everyone is a good firefighter, and everybody knows how to work an EMS call. Big deal, you got your recognition patches, you know how to do your job to get by, and you have all your book knowledge. This alone will not get you through probation.

Any department worth its salt will have some type of new hire orientation. This is where your department will show you how they do things. This can last one to six weeks, if not longer, and is usually a forty-hour work week, Monday-Friday. This will cover everything, from running Fire--EMS calls, to properly completing a report. Moreover, some departments design their orientation around a mini fire college, which includes morning workouts, live fire training and personal life saving procedures. Be sure to pay attention.

Orientation is where you need to shine, plain and simple! All eyes will be on you. Orientation is where most of the Chief's scouting reports come from. This is where station officers and crews are looking and looking hard. Think about it for a minute. If you're hired at Station 1--the centerpiece of the fire department where most station tours are given, VIPs show up and ride, including politicians--why would they want a slacker working there? The same could be said for any firehouse. Most firemen don't know you; they only know you for the fake tribal tattoo you have on your arm, even though you have never been part of a tribe. Orientation is your time to show your co-workers you belong.

You have to stand out in orientation. You have to do everything better than everyone else. Even if you're not number one, it's ok. You have to give 110% every day and maintain a positive attitude. Don't worry, that's why I'm here; I've got your back.

Follow these 10 steps, and I assure you, people will notice and say, "Hey, we want that guy."

Always help picking up the hose

10 Rules for Orientation

1. Arrive one hour early: Regardless of when your class starts or what you have planned for the day, you need to arrive one hour early, every day, during orientation. Why? People will be watching. Don't be the guy who shows up five minutes before shift starts. Arriving early is a sign of maturity, pride, and enthusiasm for your new department. It shows that you want to be there and that you are eager to start the day. Another good reason is to deal with the "What If's." What if you get a flat tire? What if there's an accident? What if? What if? What if? Remember, nobody cares that your car broke down or that traffic held you up. The instructors don't want to hear your excuses.

Besides, you will get in trouble. Don't start your career late. Show up one hour early.

2. Don't be arrogant: It's not what you did in your prior life that counts, but what you can do now. How will you contribute? Nobody cares about your prior experience, the lives you saved while risking your own, or how busy your last station was. In my old department before I went to Orlando, I was a high ranking young officer with an incredible amount of responsibility for my age. I've received numerous commendations and awards for some pretty cool stuff I did. I figured when I got to Orlando, this would instantly give me street cred with the guys. Man was I wrong! What can you do now for this station and fire department? This is what matters. If you come across as a cocky know it all, then you will definitely have a rough time. Keep your mouth shut, ears open, and be humble. Even if you don't agree with things, keep quiet. Just say, "Yes, sir", "No, sir", "Ma'am", to EVERYONE who currently works in your department, instructors, officers—everyone. Humility goes a long way. Practice it.

3. Don't talk about your family's legacy: Fire departments traditionally hire family members already on the job or recently retired. This practice maintains the brotherhood--family environment and is quite common. If you do have a relative on the job, don't expect favors or throw their name around. It is a huge turn off and frowned upon. You cannot ride on the coat tails of a past family member thinking it will help you. To be honest, the legacies have more eyes on them than anyone else. You will be compared to your relative whether you like it or not. You don't have to tell people who you are; trust me, they already know. You need to perform your duties better than your relative and strive to maintain the family's good name. Don't be a screw up and tarnish the reputation that your dad, uncle, or aunt built up over an extensive career.

4. Be the First to Start Cleanup and Re-Packing: After every evolution, or a full day of training, there will almost always be things to do such as re-filling air bottles, loading hose back on the engine, cleaning equipment etc. When done with your evolution, or after watching others complete theirs, you should immediately jump on whatever needs to be done. Don't wait and be told what to do or what needs to happen. Just do It! Be the one on top; be the person draining the line and dragging the rest of the hose to the truck. Once that's done, then jump on the bed. Do the hard tasks that nobody else wants to do. It's ok to be the one on the back step once in a while, but 95% of the time, do the hard tasks. Later on, you'll be glad you did.

5. Enjoy morning workouts: There is a good chance your department will do physical training (PT) every morning. If you're in great shape and can run a marathon, do hundreds of push-ups and sit ups without breaking a sweat, then these workouts are where you can shine. Nobody wants to be in last place or be the guy who struggles with everything and vomits all over the place (If you do vomit, don't let anyone see). Don't be the one who can't finish the run or who's whining in pain at his fifth pushup. Morning workouts are designed to develop the habit of working out regularly, staying healthy, and eating right. The key to the morning workouts--and making a good impression--first and foremost is to be in top physical shape. You want to be the person up front or pretty close to it. You want to be the one leading the charge while everyone else follows. Remember, all eyes are on you! What kind of shape is this guy in? That is what they are saying. If after working out or during an evolution, never let them see you tired. Keep your head up and don't show weakness. Moreover, Never Ever Quit! NEVER! No one wants to work with a quitter. No one wants to be deep in a warehouse fire or up in a high-rise building with someone who quits. It's ok to be tired; it's ok to grind it out to the very end, but just get the task done. Remember what I said earlier: If you are so tired where you don't think you can continue, you are probably almost done.

6. Keep Your Mouth Shut: There is nothing worse than to have somebody constantly running his mouth or trying to lead all conversations. My advice: Keep your mouth shut; talk less. You want to annoy the crap out of someone? Keep yapping away. You need to keep chatter to a minimum; it's ok to ask questions, just not one hundred questions a day, especially stupid ones. If you are sincere and truly don't understand something, go ahead and ask and have them explain it again. If you are still confused, then pull the instructor to the side after class and ask for further assistance.

Some new hire training

7. Don't Tolerate Smart Asses: Every class has one--that individual who wants to be the alpha dog, the knucklehead who is the loudest and toughest, while at the same time, puts everyone else down to make himself look good. Remember, the fire department has a pack mentality. If they see you are the weak dog then you'll be regularly targeted by your classmates. Don't tolerate this! When the alpha wannabe targets you, slam him down hard and fast! This action lets him--and others-- know you won't tolerate his attack or bad-mouthing. The others standing around know you can't be pushed around. Just don't be smug about it. Remember, you don't want to get in a fist fight; you just need him to know you can't be pushed around. Stand up to him, then let it go.

8. Don't request a Tower Truck or Heavy Rescue assignment: Another way of losing respect from your classmates and instructors is to tell them you want to be on a Tower Truck or Heavy Rescue. These units, in most departments, are highly sought after and reserved for the firefighters with several years' experience. They are very specialized units; they run all the good calls and are not as busy as the Engines or Rescues. You pretty much need to pay your dues to get on one. I will discuss this later.

9. Don't talk about your former department and how they used to do it there: You'll sound like a whiner or a complainer, and that's not the best way to endear yourself with your new co-workers.

10. Don't try to buddy up with the instructors to act like you're one of the guys: You will look like a suck up.

Proud recruits getting their station assignments

CHAPTER 12
ROOKIE IN THE FIREHOUSE

Being an effective rookie will make or break your first few years at your new department. You worked hard to get here. This part of your career is where you REALLY need to shine; there are no second chances, re-does, or repeats. If the statement, "You only get one chance to make a good first impression" holds true, it's particularly true for the rookie, especially throughout the first month.

Don't screw this up! So far, you have been advised on everything you need to be successful as a rookie, survive probation and get the respect of most of your peers. I say most, because there is always that one person who won't like you. We'll deal with that person later. This is an exciting time, and you will do great.

Here we go: Twenty-Five tips for the rookie.

1. Arrive early on your first day: If your shift starts at 8:00 a.m., then arrive at 6:30 a.m. Arrive quietly so you don't wake up the crew that's still sleeping; they were up all night. Don't piss them off your first five minutes on the job. You need to keep arriving at 6:30 until your station officer or several firefighters advise you that you don't need to show up that early. Once they see you do this for a few months, then it's ok to start arriving at 7:00 a.m. You need to show up at 7:00 a.m. for the next year and start your station cleanup; quietly. I'm so sorry you have to get up early for work, but trust me, it will be worth it.

2. Look Good: For the first few months, show up in your uniform shirt that is pressed and cleaned, especially if it's a buttoned down, collared shirt. You can take it off when someone tells you or when you're doing dirty work, but until then, keep it on. Don't show up in your issued workout clothes or civilian dress thinking you will change when you get there. Be dressed and ready to go.

3. Don't show up empty handed: Your first day, and only your first day, bring something for the crew. Show up with a bunch of bagels, donuts, or breakfast. It doesn't have to be fancy, but don't arrive empty handed. They will appreciate this, and it is a nice gesture.

4. Find the Senior Fireman: Find the senior guy or most personable one and pick his brain on what is expected of you. This is something you shouldn't have to ask, but if you find yourself wondering around, then scope someone out. You shouldn't have an issue here since most crews are pretty cool to the new guy. The officer of the station really should be the first person to meet with you and sit you down. If this doesn't happen, seek out the senior guy.

5. Make the coffee and tea: As a rookie, one of your many jobs will be to make the coffee and tea in the mornings and sometimes during the day and always a fresh batch before dinner. If it is tea you're making, get the recipe from the person you are taking the job from and ask how much sugar, water and tea bags per pot. Seems simple, I know, but don't screw up the tea. Don't worry too much about the coffee; everyone prepares theirs differently.

6. The Flag: Every station has a flag. Some are not lighted twenty-four hours a day. If this is the case, then it will need to be put up in the morning and taken down at night or when it rains. The flag needs to be folded properly; they will show you how to do this.

7. The Ice: Every firehouse in Orlando had commercial ice machines, and almost every station had an Igloo cooler filled with fresh ice in the kitchen. Fill it up when shift starts and add fresh ice before dinner. Don't forget to change the ice in your cooler on your unit every morning. Seems like a small but important matter.

8. Starting the day: Once you arrive at the station, figure out what unit you are on, or if you floated out for the shift. It's time to relieve the firefighter who's coming off duty. If you are assigned to Engine 10, as the Firefighter/Paramedic, then you want to seek out the person who holds that spot. Get a pass on from him which states what equipment is missing, broken, etc. If your shift starts at 0800, don't relieve him at 0755. That's just plain wrong, disrespectful and will give you a bad reputation. The earlier you relieve your counterpart, the better. It shows respect, dedication and a willingness to be a team player.

This is how you should prepare for your shift. Once you have relieved your counter-part, then it's time to check off the truck. I always started with my air pack, placed the mask on it and set my gear up so I was ready to go. I then moved on to the EMS equipment, replaced the batteries in my radios and Life Pack Unit. After all the EMS supplies were checked, including O2 bottle, I then would go over the fire side to go over my unit. It is 100% your responsibility to make sure you know EXACTLY where every single piece of equipment is on your unit, no excuses. If you are on a scene and somebody asks you to grab the hose clamp, KED, or quick pads, then you need to go to the exact compartment where it is located and get it. This saves time, avoids mistakes and more importantly, you don't look like an idiot.

During the day when everyone is inside watching television or talking, then you need to be outside in the bay, learning your equipment, its location and its use. The only exception to this is if you are inside the kitchen studying your SOP's and Training Manuals.

Once everything is checked on your unit, then it is time to wash it. Wash it every single shift, even if it does not need it. Keep doing this until a senior firefighter or engineer tells you not to worry about it. Only after your unit is completely checked off and washed can you help another person wash theirs.

9. Station Cleanup: Everyone usually cleans the station, but the how or when varies greatly from department to department. In Orlando, everything was assigned based on your company assignment for that particular shift. For example, in January, the Engine crew had the downstairs day room and offices. The Tower had the upstairs bathrooms and bunk rooms, and the Rescue had the Kitchen. This would rotate every month. It's actually a great system and holds people accountable for their job.

Most of the griping is usually done between shifts. It has to do with trash not taken out, dishes left in the dishwasher, sand and dirt in the engine, etc. It's OK for most of these guys to live like slobs at their own house, but not at the firehouse. If you know what you're cleaning assignment is for that month, then there is absolutely no reason whatsoever you should screw this up.

10. Do your station cleanup first: Once you know what you have been assigned to clean up, jump on it, get it done, do it well, and then help others do theirs. Always do yours first, since that is what you will be held responsible for. The rookie usually has the kitchen, along with other areas that take the longest to complete. Don't be the first one done with station cleanup. Be the last. Even if you truly are done, go figure something out to clean until everybody is done.

11. Clean the Bathrooms: Most people are pretty picky about the bathrooms. Here are some great tricks of the trade that will get your clean bathroom noticed.

The Rookie always mops the floor

A. Throw away old bars of soap in the shower and bowl. Replace with new bars but keep the wrapper on (Hotels do it this way).

B. Have extra toilet paper in the stalls and keep in package (appearance is everything).

C. If your station uses paper towels, keep plenty of them out and easy to get to.

D. Here is the best one. If you have Pine-Sol or bleach (don't mix!) then pour some in the toilets, (don't flush) and wipe down all surfaces with it. Mop floors with light bleach and water. If the bathroom smells clean, then everybody thinks it is clean. Don't go too crazy with the bleach though, you'll piss people off because now they can't breathe and start wheezing.

So, basically, what we did is not only clean a bathroom, but we wanted the world to know how well we did it. How? We just gave a firehouse bathroom a 5-star hotel royal treatment by leaving the soap and toilet paper in its original package, with a wonderful fresh scent of Pine-sol in the air. Remember, if it looks and smells clean, then it is!

12. Stay off your cell phone: I know it's hard for most people to stay away from their phones. Texting, selfies, and personnel phone calls preoccupy most people. If you absolutely have to text your mother to tell her your first day of shift is going great, then go to the restroom and do it where no one sees you. Even if you were not on your phone all day, but just that one time, if anyone sees you texting, they will think you're always texting. I know it isn't right, but that's just the way it is. If you want that important picture in front of the engine your first day, then have someone take it for you quickly, and then put your cell phone away.

13. Always Be Studying: Now that your unit is checked off, station duties completed, it's time to hit the books. Everybody else may be in the recliners or shooting the bull outside about a recent fishing trip, but not you. You need to grab all your fire department manuals they gave you when you got hired and lay them out on the kitchen table. That is where you will live until everybody goes to bed at night. Hate to tell you, but you're not going to be the first person in bed or close to it. During the day and when everything is settled down, you need to have your face in the manuals. You will be learning everything there is to know about the job. Plus, it looks good for you. You cannot be caught asleep in the recliner or in your bed early. You need to be studying or outside looking over your unit. If you're not doing this, then find something productive to do. It helps if the crew sees you do it.

If you know something has to be done, don't wait to be told what to do. If the EMS supplies just came in and they're sitting in the boxes in front of the closet, then put them away. If you are assigned to the Engine or Rescue for the day and the Tower crew is in the bay, training or going over equipment, then you need to be there next to them. Just don't tell them you hope to be assigned to the Tower as a rookie. Not a good idea.

14. The Dinner Table: If you ever want to have a great time, bond or just get pissed off at somebody, then you need to gather at the dinner table for the meals. The dinner table has a long tradition in the fire service where everybody gathers to bond and let everybody else know how smart you are (note tone of irony). Seriously, this is the place to do it. The crew would love to hear how you're a subject matter expert in everything.

The Crew at Station 1 eating together

Unfortunately, over the years, there has been a trend that I am completely against. It seems, especially with the younger generation coming through, that the crews are now eating out more. Most departments have a rule which states, you can pick up your meal anywhere, just not sit down and eat it at that restaurant. More recently, Fire Chiefs seem to be giving permission for crews to eat out.

I am definitely against this. Nothing is more important around the firehouse than to prepare and eat the meal together. The banter that flies back and forth is awesome. Some of the best stories come from the dinner table, unfortunately, not all of them good. It's also the place where the bitching comes out.

There are times the crew is just pissed off at fire department politics. This is where they vent because they feel the chief, mayor and everybody is screwing them over. They discuss some new shoot-from-the-hip rule that just came down the pike, or they grumble about how city management just screwed over the firemen with no raises, while the mayor and city manager make $200K plus a year. Yeah, this is where it all happens. This is where all of the world's problems are solved by a handful of blue-collar workers--at the dinner table.

15. Always Eat with the crew: You ALWAYS eat with the crew, even if you are vegan, gluten free, or on some other crazy restricted diet. Rarely did someone not eat with the crew. If someone didn't like us or we didn't like them, they might not eat at the table. If you are on a diet, prepare your meal separately but sit down with the crew and eat together. Just let the chef know in the beginning of the shift that you won't be eating so they don't count you in. Breakfast is usually eaten together on weekends and holidays. Lunch is usually on your own. Dinner, we all eat together.

16. Help with the prep work: Even if you don't cook (we will discuss more about this later) then you need to help the chef with the prep work. Remember, he may be cooking for four to thirty firefighters and needs help. Just go in and ask what you can do to help. Just start chopping away or doing the dishes.

17. Dinner time behavior: When the time comes to eat, be the last in line and don't take the last piece or fill your plate like a ravenous animal. Don't be the first one up for seconds, either. There is almost always enough for seconds for everyone, so don't worry, there's plenty. While you are at the dinner table and the phone rings, be the first to get up and answer it, (Hint: Sit next to the phone.) You should always be answering the phone during the day.

While eating dinner and another unit gets a run and it is not yours, then get up and wrap their plate with aluminum foil. This is a nice gesture and helps keep their food warm. If it's your call, then they will do the same for you. As soon as you are done eating dinner, be the first one up to start the dishes. In most firehouses, doing the dishes is shared responsibility from the Chief down. Everybody takes part with the dishes. Chances are when you get up and start washing, the crew will call you back to the table to sit down. That's a good thing, sit back down, enjoy the crew until they all get up together.

Be the last person to finish cleaning the dishes and kitchen. You will get help with loading the dishwasher and cleaning and drying the big items, but when it comes down to taking out the trash, wiping down the counters, mopping and all the small things, you will probably be on your own. It's ok; you're the rookie and you may get help from another person with less than a year on, but just do it and smile.

Always help in the kitchen. Even if you are not the rookie and have less than a couple years on, then you need to help. It's the right thing to do. Before you leave the station as you're coming off shift, do one last walk around the firehouse, especially the kitchen area. The oncoming shift will ALWAYS gripe about the kitchen. Just make sure no dishes are left in the sink or dish washer, coffee cups lying around, and the trash is taken out. If you did all this and they still bitch, oh well too bad, not your problem.

18. Work out with the crew: Most crews work out, and sometimes they do it together. Find out what time they work out and join them. Even if they don't, it's ok for you to. Just make sure you don't work out when everyone is cleaning or training.

19. Have a positive attitude: Always be positive, happy and don't complain about anything. You're happy to be there and life couldn't be better.

20. Dealing with the slob who leaves dishes in the sink: Just prepare yourself; it is going to happen. You just cleaned the kitchen after lunch and dinner, and that one slob decides to leave his pots, pans and dishes in the sink. This used to annoy me royally! To me, there is nothing more disrespectful than dirtying up an already clean kitchen. Here is how you deal with it. If you're new and have far less than a year, well, sorry you're out of luck. There really is nothing you can do. If you're eight to twelve months on the job and you want to mention it to the slob, then that's ok; just be polite about it.

After you're off probation, it's game on, Slobs! If after you are off probation and no longer the rookie and Mr. or Mrs. Slob continue to leave their nasty half eaten dishes in the sink, then the first thing I would do is pull them aside and speak to them nicely. If it still continues, then take those dishes and put it in their bed where they sleep! It basically lets the slob know you're not going to tolerate their behavior. It puts them on notice that you are not their slave spouse and will not tolerate it. I'm sure this will create a scene, and the station officer should tell the slob at this point to do their own dishes and quit trashing the kitchen.

If, by chance, the station officer does not get involved and Mr. or Mrs. Slob continue to leave his or her mess for you, then you need to be ready for the nuclear option. This was actually done once. Take all their dirty dishes and mail it in a box to their home. Yep! That's right! We are going nuclear here. Make sure you address it to their spouse with a letter that states.

Dear Mrs. Smith,

It seems your significant other is used to having his dishes done for him at home. Danny accidentally has been forgetting to do his own dishes at the firehouse. I have taken it upon myself to mail these dishes to him for you to do. Thank you for your understanding in this very important matter.

Sincerely,

Mauro Porcelli (Not your husband's caretaker)

Ok, so maybe instead of just off probation you should have a few years in before going nuclear, but I think you get the gist of things.

21. Wake up first in the morning: Do not sleep in! When you wake up, do a quick walk around and make sure everything is tidy. If not, tidy up before you leave.

22. End of shift pass on: Before you go home, give a good pass on to your relief. Let them know if anything is wrong with the equipment and what needs to be replaced. Do not leave anything low such as O2, your air-pack, or anything that you could have replaced yourself to the oncoming shift. Take care of it yourself and don't leave it to the next shift to handle.

23. Don't be the first to go home: If you are the first to be relieved, ask another person of equal rank if they want to be relieved instead. This shows respect to a senior member and doesn't look like you can't wait to go home. Once you are relieved and ready to go, your shift is over. It's ok to hang out with the oncoming crew and talk, but don't make yourself at home lying in the recliner watching a movie. Go home!

24. Do your reports: Reports suck, plain and simple. I hated doing them, and so did most people. They take time and almost always get rejected by administration because something is not correctly documented. You need to do all the reports that your unit responds to. Don't let them pile up. Some reports you can't do because it is not your role, but do whatever you can, but don't let the senior person do any.

25. Take the initiative: Take the initiative to do what needs to be done. Do not wait to be told what to do. If you see something left undone, then take care of it.

How you handle your rookie year is up to you, but if you follow these tips, even if some reactions seem radical, you will earn the respect of your co-workers and survive your rookie year.

CHAPTER 13
IT'S YOUR TURN TO COOK

That dreaded day has come. You skated by for weeks or months without cooking a single meal. You have been helping in the kitchen as a faithful assistant by cutting and chopping vegetables and stirring the sauce when the chef was out on a call. But you still refused to cook. Most people don't venture into the kitchen because they are either too lazy, or they say they don't know how. I'm throwing the BS flag on this one. Everyone knows how to cook something, anything. This is your chance to learn how to put something on the table that others will enjoy.

Word of warning: If you have never cooked for a group of picky people, especially those who complain all the time, then get ready for some good old-fashioned firehouse ridicule. No matter how good your food is or how much you make, some knuckle-head will always complain that it sucks, and yet, he'll go back for seconds. That's about the only way you will know if your food passes muster.

Why should you put up with this nonsense? You need to do your part. Your mother doesn't work at the firehouse to feed you. Besides, at some of the bigger stations, the cook doesn't do dishes or pay for dinner.

Here are some rules to help you manage in the firehouse kitchen.

1. Stay on budget.

We used to collect $5 from every person to cook the meal. Over time, inflation took hold, and $5 didn't go very far. Numerous times we had to collect $6 or $7 from each person. When this was done, the crew would get pissed off, moan and cry like babies because they had to cough up an extra buck. The complainers usually were the guys that never cooked anyway and their opinion never really mattered to me. What we eventually did was raise the dinner price to $8 per person. If someone was on overtime, that person would throw in an extra $20. If we had several people on OT, we would then eat a great steak dinner and took the left-over money and put in our station savings account, aka "Station Syndicate."

2. Cook whatever you want.

The worst thing you can do is ask the crew, especially a large one, "What do you want for dinner?" Each person will request something different, but no one wants to cook it. If you are the chef for the night, then cook what you want without input.

3. Cook a lot.

Whatever you are cooking, it does not need to be fancy. Just make sure there is plenty for everyone to eat a healthy portion where they can get seconds. If you are lucky, you will have enough leftovers to provide lunch for the next shift.

4. Ask for help.

Most crews are more than willing to help out with the prep work. The prep is what takes the most time; the more help you get, the better off you will be.

5. Eat at a reasonable hour.

Have dinner on the table by 5:00 or 6:00 p.m. I hated eating late. The problem is, if you're not cooking, then you don't have a say. I always had dinner ready by 6:00. That is when normal people eat, not 8:00 p.m. The only exception to this rule is if the entire station is out on a call and gets back late. Sometimes we just ordered pizza and saved the uncooked food for next shift.

6. Just cook something.

If you really are one of those people who really can't cook because you were bottled fed your whole life, then pick something simple like pasta. You can't screw that up. Just throw some store-bought marinara sauce in a pan and boil some pasta. Serve with a side of salad and you're done. Perfect. the crew will love you, and you did your part.

7. Measure portions properly.

The standard rule of 1/4 pound of meat per person is hogwash and doesn't work. Here's the best way to decide how much meat, vegetables and supplies to buy. Look at the package, decide how much you personally could eat and then multiply that by the amount of people you have to feed. I have used this technique twenty-five years and it worked for me. To be honest, if anything, I probably bought more then I needed, which is way better than not having enough. Use this strategy whenever you buy supplies for dinner. For your convenience, I have provided some tasty, simple, tried-and-true recipes to get you through your first attempts at being chef.

CHAPTER 14
COMPLAIN OR REMAIN SILENT: CHOOSE YOUR BATTLES

A very wise person once told me, "A complaint is something that truly bothers you, and you can't let it go. You feel like you have been wronged and need an explanation. A whine is a constant complainer about everything; everything sucks; nothing is ever good enough, and the whiner lets everyone to know it." The whiner could be drowning, is thrown a life preserver, but whines because it's the wrong color. In reality, nobody cares about your whines. They just annoy everyone.

I learned early on that if you act professionally, have a good attitude, and are positive, then you can pretty much get what you want. I admit, getting what you want can be difficult at times, but cooperating will bring you rewards later on. For example, over the years, my Chiefs and officers knew that I was a low maintenance employee who would go with the flow. I didn't complain much, and the older I got, the less I debated.

I wanted to save my energy for the important stuff. They also knew that if I went to them with a complaint, chances were good it was something significant that really bothered me and warranted their immediate attention. More likely than not, the Chiefs took pretty good care of me. I didn't cause any grief. I was professional and didn't ask for much. I am grateful for their consideration. The whiner, on the other hand, won't get a thing but rejection at every turn. This is not always the case because sometimes the squeaky wheel gets the oil, but in more instances then not, he gets rejection, or worse, ridicule.

Lt. Ferron Horn was just not putting up with it

You have to pick your battles. You cannot fight every little thing that comes along. You will lose, develop a bad attitude, and become disgruntled. It is a horrible and lonely way to go through your career. On the other hand, if you are right, and you feel you have been wronged or treated unfairly, then stand your ground and let your words and feelings be known. Over the course of your career, you will be falsely accused or harassed. If you are correct, and you sincerely believe you are, then you need to speak up; do not back down.

For example, I was assigned to Tower 10 with a new guy. He was only on for a couple of years when he thought he should be with me on a rotation on the Tower Truck. His argument was that he, too, had the courses and should ride on the Truck, while I floated on the Rescue for the shift. I called Bullshit on this! What this guy seemed to have forgotten is he and forty-six other firefighters almost got laid off just a couple years earlier. Due to a vote that the entire department had to take, we gave up our raises so that these forty-six firemen could keep their jobs.

What pissed me off is, not that he wanted to be on the Tower Truck, but that he and his Lieutenant completely went behind mine and my Lieutenant's backs and took the idea straight to the District Chief at the time, who thought it was a good idea. I called for a special meeting with my Lieutenant, the District and me. I pled my case. The Chief did not agree with me and advised that I would be rotating with the other firefighter.

WRONG! No way was I going to take this lying down. I respectfully advised the chief that I would not accept his decision to rotate me and that I would take the dispute as far as I needed to, if necessary. I was not going to be treated this way.

Things quickly got heated. The Chief advised me that, while he was actually extremely agitated and pissed off, that I would even question him and his authority, he looked me dead in my face and said, "I tell you what, Porcelli. You don't want to rotate, then I will float you anywhere in the city, every shift that a Paramedic is needed."

Wrong thing to say to me! I then got up, pointed my finger straight back at him and stated, "I don't know who you think you are, but you do not threaten me like that!" My Lieutenant, who was sitting next to me at the time, was kicking me under the table to calm down. Let me tell you, I was not taking this one lying down and was ready to go all the way to the top with it. Was I insubordinate? Oh, heck yeah, but I will not be threatened like that by anybody, especially since I am a low maintenance employee and never gave anybody any problems. But I'll be damned if I was going to give some young, punk kid my spot after I voted to save his job. So, what happened after this altercation? I never rotated out of the station, and I never left the Tower.

Sometimes it pays to know when to keep your mouth shut in order to save your reputation for the serious battles. Keep in mind every battle must come to an end. Sometime you win, and sometimes you lose. Know when to throw in the flag, if you are right, but also know when to avoid the fight. Just remember, pick your battles and don't be a whiner.

CHAPTER 15
HOW TO GET SPECIAL ASSIGNMENTS

In most departments Special Teams, Tower, Heavy Rescue, and Haz-Mat are the prized units by the new guys. These can be very difficult units to be assigned to because there are so few of them with so many wanting a spot. Here is your best chance of getting on one.

1. Have all your courses for that unit. Example, if you want to be on a Tower or Heavy Rescue, then you need to get "Tech'ed Out" Basically, you need to have every single specialized class that the units require such as Ropes Technician, Collapse Tech, Trench Tech, Extrication Tech, Confined Space Tech, and Extrication Tech. These are just a few, and every department and state may have different requirements and courses. If you want on a Haz-Mat unit, then get your Haz-Mat courses. Most of the time this is done on your own time, but most departments will pay for it with your educational reimbursement.

2. Be assigned to a station with one of these units. It is much easier to be assigned to a special ops unit if one is already assigned to your station. The reason for this is, when a spot becomes available, the Chief Officer will usually go to the crew and ask who they would like to fill that spot. Not always, though. Most of the time, they pick a person who is in-house. They already know this person is a good guy, gets along well with others, is not lazy, trains regularly with the crew and already knows the truck. He occasionally will float over to cover a spot. The benefits of working at a station that has one of these units is priceless, especially if that is your goal. Obviously, it is out of your control where you work, but if possible, try to get reassigned to one of the stations that has one.

Orlando Tower 1 and Tower 2 working it on a commercial job

3. Stay in Shape. You can't be a slob and get assigned to one of these special units. Your chances are about zero. Most of the crews on these units are buff, Type A personalities and usually go above and beyond with their fitness programs. Get on good terms with the Officer on the unit and Chief. Hey, it doesn't hurt to suck up a little, but don't be obvious where everyone thinks you're married. Just be personable, give good firm hand shakes, not the wet wash rag kind and be yourself. Lots of firefighters tend to not say much to the Chief for whatever reason. Most of them are pretty good people, just trying to do their job. Be nice, show enthusiasm for your job, and you may get a shot at one of those Trucks.

Lt. "Gonzo" and me on Tower 10 during a massive commercial job

4. Have a good positive attitude. For the most part, these people usually don't complain and are usually professional around the Chief. They are experts at playing the political game and know how to smile when the right people are around. To be honest, if you have to smile a little bit when the right people are around, then that's ok. It's better than having a frown and being miserable.

CHAPTER 16
ASSIGNED TO A STATION YOU DON'T LIKE AND OTHER STATION ISSUES

Sooner or later, you will be assigned to a station that you hate. There are so many reasons you might hate it: Too busy, too slow, don't like the area, don't like your crew, bad vibes, or whatever. Whatever the reason, if you are the rookie, too bad, you'll have to suck it up. Don't tell anyone you hate your assignment. Your life will be more difficult if this gets out. On the other hand, if you have been on for a few years and have a good relationship with your officers, then, if you put in for a transfer, chances are, you probably will get it.

If not, then you need to remember this. The fire department runs in two to three- year cycles. Wait a bit, and in a few months or year, ask again. It has been like that since the day I got hired in 1988 and stayed that way throughout my entire career. If you are at a station you don't like, and no one helps you transfer out, then you're going to have to weather their cycle storm. Major transfers seem to occur every two to three years.

Similarly, major fire department scandals occur every two to three years. There is always something that comes up in this time frame that might provide an opportunity you didn't see coming. One of the worst things for one's career is being assigned to a place that isn't a good fit for you. Unfortunately, there is nothing you can do about it if you can't get transferred. Keep your head up, remain positive, let everyone know you want out, put in a written transfer request, and hope for the best. Within a few years, better assignments open up.

A. How to get what you want. In order to get what you want in the fire department, you need to build credibility with your peers, supervisors and chief officers. This is usually a long and subtle process. First, know your job, do your job and try keeping your mouth shut whenever possible. Your crew will notice, your chiefs will notice, and your overall reputation will be favorable. Word of caution: There is a difference between building credibility and kissing ass.

There are several subtle actions that will help you get what you want. Always acknowledge a supervisor who walks into the room. Do this with a good firm handshake and greeting. It's fine to make small talk with everyone. I always enjoyed this part. I liked to talk to anyone and really appreciated most of my supervisors anyway. Most never gave me a reason not to like them. They usually took care of me and respected me back. So why not reciprocate? Better yet, if you're cooking, invite your district chief over for supper. You may get some good inside information on what's going on around the department. Also, most Chiefs truly appreciate the invitation. Besides, you never know when you will need that favor. It's just good sense to extend courtesies to the supervisors.

B. Dealing with bad co-workers. The only thing worse than being at a station you don't like is working with people you don't like. You need to understand something: No matter how nice you are, how politely you treat your co-workers, and how mild- mannered you are, some people won't like you, plain and simple. Rarely will you dislike the entire crew, but almost always there is that one shit head you can't stand.

This is where a large department offers a solution. In a house with fifteen to twenty guys, it's a lot easier to deal with that one guy than a house with just four. With only four workers, there is no escape. With twenty, you can usually avoid that person. Unfortunately, this is not always the case. The co-worker could just be a bona fide asshole. He'll find you, but it will take him longer in the larger department.

Never accept anyone's abusive behavior. No one has the right to do this to you, not even your officer. These are potentially harmful situations that must be addressed as soon as possible, or else your life will be a living hell. If the bad behavior becomes unbearable, then you must act. If it's an equal and you can't resolve the situation among yourselves, then you need to seek remedy with your station officer. If you don't, then the matter will only escalate. If you cannot get a grip on it, even after your officer is involved, then you need to put in for a transfer. Don't tough it out, hoping things will improve on their own, because they won't.

C. Dealing with a bad officer. Most officers I worked with were great people and fun to be around. Unfortunately, some had no business wearing a gold badge and white shirt. They were bad at their jobs, horrible people and stupid, at best. They lacked basic competence, people skills and professionalism and were a total embarrassment to their positions. These individuals lacked respect for themselves and their crews and had no idea how to lead.

So, you ask, "How did they become officers?" Simple, they passed the promotional exam or gave some other favor to the chief. More than likely, he kissed ass while selling his soul to the devil. This may seem harsh, but many of us wondered the same thing and marveled at how inadequately some of these officers performed their duties. Fortunately, the fire department does not have many of these people around. Rest assured, every department has them. The sad thing is, some of these officers could be effective, but they have little to no people skills, and their narcissistic behavior will not allow them to play well with others.

So, what do you do if you have to work for one of these clowns? Unfortunately, you are going to have to ride it out. These people don't last. They are fly- by-night and go through crews like dirty laundry. Because they are not respected, they eventually burn their own bridges with everyone. So basically, try and stay out of their way; let him or her self-destruct. They eventually will move on.

A dirty little secret in the fire department is you need to have your very own " Little Black Book." Keep this book at home. In it, you will document every little detail that this incompetent officer is doing to you and others. I assure you, the incompetent has one too, and is keeping one on you. If the unfortunate day comes when you are forced to defend yourself, then you have the proper documentation to defend yourself effectively. Like I said, these are rare events, and harassment should not be tolerated by anyone. If it becomes unbearable, then go up the chain of command with your documentation and move on.

D. My 3-day rule. If you really want to screw up your career, fire off an angry email and send it to your superiors. While you're at it, cc it to the entire department. Do Not do this! Use the three-day rule. Throughout your career, there will be times that you will be boiling mad with management. It doesn't matter what precipitates the anger. What matters, is that you are angry beyond control. If you feel the need to say what is on your mind, type out your letter, but do not hit the send button for three days. You will realize that 99 percent of the time, after three days, the letter was too strong, too personal, inappropriate, and counter-productive. After three days, you will calm down and either revise the letter, or, better yet, delete it.

Renee Bishop; Left and Lt. Lori McDonald hanging out after a fire

CHAPTER 17
YOU HATE YOUR DEPARTMENT AND OTHER BURNING ISSUES

If you find yourself in a bad situation, then you have to move on. You have to find another department and do whatever you have to do to get out of your misery. This is actually quite common. There are countless reasons for your unhappiness--bad management, poor pay, difficult co-workers, all sorts of reasons. If you are in this position, then the earlier in your career you can make the move the better. You really don't want to be in your department for fifteen years when you decide it's time to move on. You will lose seniority, pay and all sorts of benefits.

Decide within the first five years of your career if this is what you want to do and where you want to do it. Most firefighters have worked one or two departments before settling in to their "permanent" home. Don't just apply to your favorite place; you need to inundate your desired area with applications until somebody hires you. If you make the decision to move on, then any other department is an improvement on where you are currently employed. Go ahead and make the move.

Dealing with the Union

Let's get this out of the way right now. If your department has a union--chances are that it does--then you need to be part of it, no questions asked. I don't care, and I can assure you nobody else does either, what your political affiliation is or if you are pro or against unions. Whether a new hire or a veteran fireman, you MUST be a member of this organization. We all have our opinions, pro, against, neutral, whatever. Just remember, the union is there to help you; the organization represents you. Everything you have today is because of the past and current members of the union. You have pay, benefits, safety and personnel management because of their work. Don't say you can't afford the dues. You wouldn't even have the salary you do have, without the work they did for you when they negotiated your salary.

In the beginning of my career, before I worked with the City of Orlando Fire Department, I was with a small Florida County department. We did not have a union, and man did it show. It was a hodge-podge department back then, with bad management, ineffective leadership, inadequate equipment, and no safety protection. It was a fire department nightmare.

People would get reprimanded or fired for stupid reasons. Once we voted for the union, good things started happening. Because of the union, that fire department is a great place to work now, with state- of- the- art equipment and training. Thanks to the union's efforts, current firemen can enjoy protections and a decent wage.

When I started with Orlando FD, I noticed immediately that we had a strong, well established union. We had the best equipment, best training, awesome SOP's and a road map on how to handle almost everything, from discipline to vacations. When you have a strong, vibrant union, everything else falls into place and makes the organization great. If you pay your employees well, give them great benefits, a good working environment with a road map to a successful career, then you attract a better worker. You attract the brightest and strongest candidates that a pool of employees has to offer.

This is the secret sauce that many workplaces don't seem to understand. Don't get me wrong. In Orlando, we had our issues, and so did everyone else, but they were fewer and less disastrous. Or, they only occurred in that two to three-year cycle I mentioned earlier. Some non-union departments seem to be in the trauma cycle forever. There are some good, non-union departments, but they are few and far between.

Your union is only as strong as its members. You need to be involved as much as possible. If you decide to stick to your guns and be that guy who doesn't join the union, then good luck with your firehouse life. No one will speak with you, do time trades with you, or have your back with anything. You have decided to stand by yourself and be the outcast. This isn't an advisable place to be in a profession that survives on team work.

It's ok to disagree with your local union on issues. I, for one, did not agree on everything they did or represented, but I know they always had my back. The few times I did not agree with certain issues, I would pull the union president aside and have a private gentleman's conversation with him. This was all it took to get my ideas presented.

Don't be a jerk about disagreeing with union representatives. The worst thing you could do would be call out the Union Executive Board in an open forum. You immediately put everyone on the defensive, and you won't get what you want. They are great people. Be smart about union confrontations and disagreements.

How to handle being the brunt of jokes

If you are going to do this job, then you'll be the brunt of jokes. You need a thick skin, or else you will have a difficult time. Keep in mind, everyone takes turns wearing the dunce cap. It's part of the job. Most of the time, the verbal attacks take place at the dinner table or while hanging out behind the station after dinner. It's going to happen. Prepare yourself.

How do you protect yourself? Just make sure you get your shots in when it's someone else's turn being the target. Everybody handles this differently. To be honest, most people tend to bring the attention on themselves by saying or doing stupid things or not thinking before speaking. They constantly express their political opinions or interject themselves when they should sit quietly. Some people can't help themselves; they just don't know when to keep their mouths shut.

So, here's my suggestion on how to handle it. If you find yourself being a punching bag, just remain silent and smile. Believe me, it works. The more you try to defend yourself, the dumber you are going to sound, and you will give them more ammo against you. Just smile, keep your thoughts to yourself; eventually, they will run out of steam.

Never forget your gear at the Station when you go on a call

Five ways to keep the crew from stockpiling ammo against you:

1. Keep your political opinions to yourself, especially if your opinion goes against the grain. You won't convert them.

2. Keep your girlfriend at home. Nobody wants to see your wife, boyfriend, girlfriend at the station all the time. And, sure as hell, don't invite her for dinner. If you want a visit once in a while, that is fine, but not every shift. When she does come over, limit the visit to thirty minutes, no more.

3. Leave your personal life at home. Nobody cares how good you have it or the tough time you're going through because your wife is having an affair. Well, they, actually do care about that, but only because they can use it against you at the dinner table. They may get a good laugh out of it or try to hit on her themselves because she seems pretty easy. If you do want to tell someone about home issues, do it in private and not in a large group.

Make sure your girlfriend doesn't hang out all day

4. Don't tell anybody about how rich you are or that you have a trust fund. That is private information, and the firemen don't need to know.

5. If you happen to be tight with the Chief or upper management, keep that to yourself. They will view you as an ass-kisser, and you will lose credibility and respect. Keep any relationship you have with your superiors to yourself. All of this appears like you are bragging. No one likes a braggart. Keep your private matters and political opinions to yourself, and you won't be the target of jokes.

CHAPTER 18
FOUR KINDS OF
FIREHOUSE ASSOCIATES.

In my department we had five hundred firefighters. Throughout my career, I had about four hundred-seventy co-workers, twenty friends, but only five true friends. Of the many co-workers, I encountered only five individuals that I truly couldn't stand. When you recognize who's who in these categories, your career will go a lot smoother. Unfortunately, this takes time, sometimes years. Certain experiences help prepare you for each type of individual you will encounter. All co-workers will fall into one of these four categories.

1. Co-Worker. This is the person you work with that doesn't give you any problems. He or she is there to lend a helping hand, generally to have good casual conversation with, and to mind his own business. He's or she is usually a good person, but you wouldn't want to go on vacations with them, not because they're jerks, but because you don't have much in common.

The women of Orlando Fire Department hanging out on their day off

2. Friend. This is the person you will go to fire department functions with such as campouts, sporting events or group functions. This person is usually pretty cool and fun to be around. He makes you laugh and is there for you when you need him. I wouldn't tell him where the bodies are buried or ask for advice on marital issues, but I would vent to him on firehouse bull crap or ask for special favors.

3. True Friend. Now these people are rare. These are the ones you can say anything to. They have always been there and will always be there for you. These are the people you would go on family vacations with and share confidences that you know will not get spread around. These are the people that are there after you retire and will stay with you until you die. We have always said around the firehouse, "There are 500 firefighters, but you will only have enough true friends to count on one hand".

4. The Sellout. The sellout scumbag piece of crap! This is the person you avoid. If they died you wouldn't shed a tear, definitely not go to the funeral, and would never help on the side of the road. You probably hope there was a puddle of water while you are driving by, so you could aim for it and soak him.

This may sound a little harsh, but, unfortunately, you will encounter these people, the lowest of the low. These are the ones that pretend to be your friend, sometimes a best friend, and then decide to sell you out over a promotion, a unit assignment or to look good in front of his other buddies at your expense. These people are toxic and hazardous to the organization. If you see them harm or discredit a co-worker, chances are good they will do the same to you, too. They will sell you out in a heartbeat to management and act as the fire department Hall Monitor. Stay away from these dirt bags. Once you have been sold out-- and you will—neutralize them immediately. Let them know in the presence of everyone else that you know what they did. This action puts a warning sign up to others to stay away from their toxic personalities. I think you get the point.

Burnt on the outside, raw on the inside, with my buddy, Lt.
Dave Williams Orlando Fire department (Ret)

CHAPTER 19
OFF-DUTY TIME--DON'T BE AN IDIOT

One of the great things about the fire department is all of the fun adventures, vacations and partying that you will do with your fellow firefighters. There is nothing better than to hang out with your best buddies away from the job, drinking, laughing and having a great time. The laughs, antics and stupid moments will last a lifetime. You will see and do things you never imagined and didn't think possible. You will wonder afterwards, how you survived and not end up in jail. You will see your single buddies buy drinks all night for some hot looking girl, thinking she was really into him. Then, you find out she just wanted free drinks, is a prostitute or left the bar with your best friend who didn't spend a dime on her. You will go to the firefighter state games, stay out until 5:00 a.m. and then have to be awake by 7:00 a.m. to make your golf tee time. You and all your buddies are hung over; there are people in the room you don't recognize, and yet, you do it all over again the next day. These shared experiences don't fade.

It is common for the rookie or veteran firefighter to say that it is not the department's business what I do on my days off. This statement is true for the most part. Your private life is not their concern. The department's primary concern, is that your off-duty life does not reflect badly on the organization. If you want to act like a complete jackass, don't get caught. Whether you accept this truth or not, you are a direct reflection of your department and the uniform you wear.

With social media today, there are no secrets and no expectations of privacy. There is nothing worse for a fire department than to see one of its firefighter's mug shot plastered on front page of the newspaper or on six 'clock news. It doesn't matter how minor of an offense, it looks bad, and it certainly doesn't help your career. The public generally doesn't care if a regular citizen is arrested for a DUI, hires a prostitute, or is frequently publicly intoxicated.

But when you are a fireman, and this happens, all the local networks will cover it, you will be shamed, not to mention the brunt of jokes for years to come. Your mug shot will be taken off of the internet, photo shopped, and hundreds of copies taped all over the fire station. This will greatly irritate you, but there is nothing you can do about it. I've warned many times already: Everything you do adds up to reputation and behavior. If you do something stupid, suck it up, because you probably deserve the fallout.

Depending on which department you work for, everyone will have a different policy on how to deal with inappropriate conduct. In the Orlando Fire Department, if we were arrested for any reason and taken to jail, we had twenty-four hours to contact the chief or his designee. Some departments are more liberal than others. We had it pretty good in Orlando and were considered innocent until proven guilty.

Unfortunately, many places are not that tolerant; they terminate your employment first, ask questions later. You then have to fight to get your job back, while possibly spending thousands of dollars in attorney's fees, not to mention your court fees. If you are found guilty-- correctly so—and you feel your career is at a crossroad, or that a hefty suspension is coming, it is best to admit your guilt, plead for forgiveness, take your punishment, and move on. As with any disciplinary action, contact your union representative and seek their advice.

Just remember, you are not above the department and you will be held accountable for your actions. This includes on and off duty.

Ice Hockey Team celebrating a victory during the Florida Firefighter Games

CHAPTER 20
FEMALE CO-WORKERS

We might as well address the issue of female co-workers in the firehouse. Firefighting, traditionally a male-dominant career, must now adjust to increasing numbers of females in the workplace. Men together, in the absence of women, tend to strut around, flex their muscles, utter profanities, and argue vociferously about insignificant matters. It comes as no surprise that the firehouse is not known for political correctness. With this in mind, when you join the fire department, you have to accept the current realities of the firehouse. Even though women comprise only seven percent of firefighters, according to the National Fire Protection Association, that number is likely to increase, and you will eventually work with a female co-worker.

What do male firefighters need to know regarding female co-workers in the firehouse? First of all, be respectful! Yes, you have a first amendment right to freedom of speech, but that right does not include the use of language that could be construed to include sexual harassment or suggestive language. Men often get confused by this. What a male may consider as harmless banter, the female might consider offensive and demeaning. The next thing he knows, he is called before a committee to fight for his job.

But let's be honest, the male bravado, machismo, chauvinism and lack of respect towards females can be a problem anywhere but especially in a firehouse. I have seen many of my male counterparts get reprimanded, demoted, even terminated because of sexual harassment. This is all avoidable if you show courtesy and respect toward your female counterpart, or anyone, for that matter. Remember, there is a line you cannot cross.

Where is that line? Here is one definition of sexual harassment from the US Equal Employment Opportunity Commission:

"It is unlawful to harass a person (an applicant or employee) because of that person's sex. Harassment can include "sexual harassment" or unwelcome sexual advances, requests for sexual favors, and other verbal or physical harassment of a sexual nature.

Harassment does not have to be of a sexual nature, however, and can include offensive remarks about a person's sex. For example, it is illegal to harass a woman by making offensive comments about women in general. Both victim and the harasser can be either a woman or a man, and the victim and harasser can be the same sex. Although the law doesn't prohibit simple teasing, offhand comments, or isolated incidents that are not very serious, harassment is illegal when it is so frequent or severe that it creates a hostile or offensive work environment or when it results in an adverse employment decision (such as the victim being fired or demoted). The harasser can be the victim's supervisor, a supervisor in another area, a co-worker, or someone who is not an employee of the employer, such as a client or customer." The harassment could be inadvertent in the eyes of the perpetrator but demeaning to the victim. This is why firefighters need to develop a new sensitivity to the seriousness of the issue.

Here's another clarification of sexual harassment:

The Equality Act of 2010 uses this definition for sexual harassment: "Any unwanted conduct of a sexual nature which has the purpose or effect of violating someone's dignity, or creating an intimidating, hostile, degrading, humiliating or offensive environment for them. It covers indecent or suggestive remarks, unwanted touching, requests or demands for sex and the dissemination of pornography."

When you read these definitions of sexual harassment, you realize how easily someone can be accused of inappropriate conduct. You must be responsible for your actions and avoid any and all appearances of harassment.

One area might be particularly problematic for firemen. At most firehouses, the dorm room is a common area where both males and females sleep, sometimes sharing the same restroom. Use common sense here. In these situations, you must take all precautions necessary not to put yourself in a situation that can be deemed offensive. Here is a good example. Let's just say you have that common dorm room with a shared bathroom. You took a shower and walked out of the bathroom wrapped in a towel. None of your private parts are exposed, but you have to pass a female's bed to get to yours and get dressed.

What would seem like an innocent act can actually cause trouble for you. Prior to taking a shower you need to advise your female coworker that you will be taking a shower and that she may want to wait to lay in her bed until you're done. Or, take your clothes with you for your shower, get dressed there, then go to your area. Either way, you are making the effort not to cross a line. The bottom line is to mitigate any possible situations that you may be in. This will reduce your exposure to accusations. Always be respectful of others and never put yourself in a situation you will regret.

CHAPTER 21
DON'T LOSE THE JOB YOU LOVE

Sexual harassment is not the only potentially career-ending behavior. Here are a few that will get you demoted or fired. Most people actually do their job quite well. Once you pass probation, don't think you are home free. Too many firefighters get terminated years after their probation ended. Unfortunately, most of the time, termination was avoidable, even after the employee had several warnings. In the case where the employee didn't deserve firing, the union will fight for the firefighter's job. Here are some of the main reasons firefighters get fired.

1. You test positive on a drug test: Firefighters losing their job due to a positive drug test is actually not uncommon. They either get called for a random drug test, wrecked the truck and had to get tested, or got hurt and were tested. We all know when we get hired that we can and will be tested. Some departments have a zero-tolerance policy, while others make you go through rehab and then random you for a year or two afterwards. Do yourself a favor and stay off the stuff entirely.

2. You're Caught Lying: We in Orlando had a saying. "Never lie about what you did." If you did something wrong and get busted, then just fess up, take your medicine and move on, especially if it is a small violation. You will probably get your hand slapped anyway.

3. You're Caught Stealing: At the firehouse, you can leave a $20 bill on the table for a year and no one will touch it. If you leave a Klondike Ice Cream Bar in the freezer, chances are it will be gone in minutes. You will piss off more people by stealing the Klondike than stealing the $20 bill. At the firehouse, especially in our lockers which are usually open, we have our wallets and all sorts of personal items lying around and open for all to see. I don't think I ever had anything stolen. But I can assure you that if you did steal something, then you're done. No one wants a thief around.

4. You're Late to work: Don't be freaking late! You know you have to be at work at 0800. Don't show up five minutes before your shift. First, you're going to piss off your relief because they want to go home and don't want to risk a late call. During your career, you might be late a few times, which is no big deal. The guys that you always relieved at 0715 or 0730 would be happy to hold over for you. Plus, the station officer probably won't say anything because he knows you're usually reliable. But if you're that person that's a 0755 guy, then you're done, and nobody will help you.

5. You Access Internet Porn: Look at this stuff on your own time. Every department has computers. With this computer, you have a personal ID and password. They will track those who go to inappropriate web sites. You will get caught. Just don't do it.

6. You Make Racially Charged Remarks: Another quick way to get fired is to get into a racial argument with the other race and start using racial slurs. Be careful with this because the definition of being a raciest is, whatever the other party deems offensive.

7. You Make Anti- Gay, LGTBQ Comments: This is definitely something that is new in the firehouse. In my day, we just didn't talk about homosexuality. If they were in our midst, they stayed in the closet because of the stigma attached to it. We would make gay jokes because most firefighters didn't know anyone, especially another firefighter, who was gay. There were probably a few in the department, but we just did not know. Times have changed; homosexuality has become more open. Most people, today, don't think twice about working with a gay partner. Sexuality is an individual's business, not mine, and shouldn't be yours either. Don't make any gay jokes around the firehouse; the person next to you may be and you just don't know it.

8. You Subject Females to Sexual Harassment: We addressed this issue in a previous chapter, but it bears repeating here. If you want to get fired fast, sexually harass a co-worker by touching or making inappropriate comments, especially if you are her supervisor. A Human Resource Director once told me the definition of sexual harassment is anything the female thinks is harassment. My next question to the HR director was, "What if I thought she was pretty and wanted to ask her out?" The HR directed stated, "Make sure you're not her supervisor, don't work with her directly, and you can only ask her out one time. If she says, 'No,' after that, your continued attention could be considered harassment."

Something else to think about: If you work with a female, the typical sexual guy talk needs to stop. Once again, always be respectful around others.

9. You're Preoccupied with Social Media: This issue gets people in trouble. What you do on your own time, well, is actually not your time. You represent your department. Do not post stupid comments and photos online. If you don't want your mother to see it, then don't post it. Such behaviors as smoking a joint, being highly intoxicated with a bunch of underage girls, or any other questionable behavior will get you in trouble.

Be smart; your social media accounts are monitored. If they are not, then one of your sell-out buddies might conveniently get the pictures or videos to the chief's office. He wants that special assignment or promotion. This is also true with your Twitter accountant. You do have a right to free speech and can say whatever you want. If you're not going to post yourself blowing out a birthday candles or swimming with your new puppy, then don't post it.

10. You Have a HIPPA Violation: This is a regulation designed to protect personal medical information. You may want to take a picture of that patient on the road just thrown out of a car window, or the body run over by a train, or a compound fracture with the bone sticking out, or a passed out drunk in the road, or the crack addict getting another fix. These pictures will probably get you the hundreds of likes on social media, but they will also get you fired. Keep this stuff confidential and off the internet.

11. You Have a Bad Attitude: Having a bad attitude will also get you terminated. I am not talking about the typical bitching that usually goes on around the firehouse. I'm talking about the highly toxic, insubordinate person who doesn't follow my three-day rule. This is the person that hates everyone and life in general. You may be the best firefighter in the world, but if you have a nasty attitude, no one will recognize anything good that you do. Try and be happy. It is not that difficult.

12. You're Caught Fighting: Do not get into a fist fight at the firehouse. Someone, or maybe everyone, will piss you off so badly that you want to knock the shit out of them. Step back, relax, and chill out. This is a must because if you hit someone, you probably will get fired.

13. You're Insubordinate: No matter how mad you get at your officer, you need to keep your feelings to yourself. I'm not saying you always have to remain silent. You can speak your mind, but you cannot do it while cursing at your officer, especially on a call. You might not get fired unless you are on probation, but you will most certainly be written up. If this happens too often, find another line of work.

14. You're Caught Having Sex at the Station: It's pretty simple, just don't engage in sexual activity at the firehouse, especially on the hose bed of your new fire engine. Chances are, your buddies are probably filming you and will post it all over social media.

15. Don't Get a DUI: Firefighters love to party and sometimes party too much. In today's society, there is absolutely no reason anyone should get a DUI. With such companies as Uber and Lyft, no one should risk ruining his record and reputation by driving after drinking and partying. Use these services to find a ride home or back to your hotel, even if you think you are sober enough to drive.

A word of warning here: Firefighters tend to think that a short nap will cure the hangover. Not so. When they have to work the next day and drink the night before, their system retains the alcohol. They wake up feeling good and go to work. Once at work, their officer or co-worker notices alcohol on their breath, and they get sent to be tested. It would be worse, if they wreck the truck and hurt somebody. If you are nursing a hangover, don't come to work. Call in sick and save yourself from losing a good job.

What to do if you get disciplined

Very few people make it through their entire career without some type of discipline. Disciplinary actions range from a verbal or written warning, suspension, or termination. Most people who have been disciplined probably deserved it and had several warnings prior to the disciplinary beatdown. My rule is simple: choose your battle. If you know you're busted, then admit what you did, ask for forgiveness and learn from it. Most of the time this goes a long way.

If you are wrongly accused or have been the target of a rogue officer, and everything is on the line, then you have to fight it. The first thing you need to do is contact your union and start the appeals process. This varies greatly from department to department and should not be taken lightly. This is when you produce your little black book, that tell-all book you kept at your house. Now is the time to use it to your advantage. If you are getting reprimanded, especially if it is over numerous missteps, then your supervisor will have notes. The only way you can counteract this is with your own notes. This should include, but not be limited to, dates, times and people in the room. It could be your only saving evidence.

Final tips on how to protect yourself and your livelihood. Know your union contract. You will be surprised at how often management breaks the union contract as they go after someone. If you know your contract, then you will know your rights. Management must follow the contract. Keep your mouth shut and don't sell anybody out to save yourself. If you're busted, take the hit, but don't take anyone down with you. Most departments have a progressive discipline that starts with a verbal warning and ends with a termination hearing. If you are targeted, this progression can go quickly. This is especially true if your supervisor hates you. Do not take this lightly. Your livelihood could be at stake. Be smart about your own behavior, but defend yourself to the end if you are falsely accused.

CHAPTER 22
BEING A TEAM PLAYER

The fire department can be accused of being a group think or having a pack mentality. This might be because of peer pressure. No one wants to be a standout. There are times where you need to go with the flow and other times where you need to step back and just shake your head. Don't feel compelled to agree with everything. Sometimes, it's good to speak up, especially if it directly affects you and goes against your morals and beliefs. Here are some ideas to help you become a team player.

1. Tattoos: You just got that dream job or unit assignment that you always wanted. What now? Get a tattoo. No, don't. One of the stupidest things you can do as a firefighter, especially a new one, is to get a tattoo with your station number and or unit assignment tattooed on your arm, back, chest, lower back or wherever. A dude should never get a tattoo on his lower back. Remember my two-to-three- year rule? Almost everything changes, especially station assignments every few years. Why would you possibly put your temporary assignment on your body with a permanent tattoo? If you want a tattoo of your pet puppy, or a tribal tattoo that you don't belong to, then that's up to you. Just leave off the unit and station assignments. You'll thank me later.

2. Compete on a Fire Department sports team: One of the most rewarding ways you can bond as a unit is to hang out with your FD friends on your day off, especially competing on a team in a sporting event. Some of the best times I ever had with the squad was playing on the fire department's Ice Hockey Team. We traveled around the state of Florida and a few times to Canada to compete. Additionally, each year, we had the "First Responder Games," formally called the "Firefighter Olympics." They include every sport imaginable, and you can compete in as many events as you want. I usually did ice hockey and golf. These events promote camaraderie and brotherhood. If you really want to promote yourself and meet some wonderful people, then join a fire department team.

Guns and Hoses charity game, Orlando Arena

3. Take a road trip: If sports are not you're thing, then take a regular trip with your friends. Don't worry about your wife or girlfriend. They will probably appreciate the break from you as well. These trips were some of the best experiences I ever had with the fire department. For over ten years, a group of my best friends and I would go to my cabin in the mountains of Murphy, North Carolina. We participated in a golf event very year. The best part about it was that we played, ate and drank our hearts out. We designated cooks for breakfast and dinner, and we ate some of the best meals ever. We would harass each other until someone got pissed off or walked away. This was a trip that we all looked forward to.

Doing some off-shore fishing with the guys

Another memorable time took place in the Keys. Ten firefighters and I went on a lobster/fishing trip in the Florida Keys during lobster season. We rented a home on the water, split the cost for everything, and had a blast. For the entire week, half the crew fish while the other half went lobstering. When we got back in the afternoon, we swam in the pool, drank lots of liquor and grilled steak, lobster and Mahi Mahi every night. Do yourself a favor and go on some getaways. You won't regret it.

4. Doing time trades: I loved doing time trades and did them regularly. The fire department is one of the very few occupations where you can call your buddy and have him cover your twenty-four- hour shift. In return, you work for him when he needs you. It's actually a great benefit to all. You can swap a twenty-four- hour shift yet not use vacation time. But like other aspects of the fire department, firemen can screw this up. Most departments have a rule that no one can work more than forty-eight hours in a row without having twenty-four hours off.

This is a good rule for safety reasons. The problem arises when you take the swap but the other guy doesn't repay the swap back in a timely manner. If you don't repay the swap, you will get a bad reputation and screw yourself over because nobody will work for you anymore. With this in mind, do not forget to show up for the other guy.

In my department the other person who did the swap will get written up and be docked vacation time. Why? Because they were the ones scheduled to be there in the first place. Remember, always pay back your swaps in a timely manner and you will have a group of reliable people that will be happy to time trade with you.

CHAPTER 23
SICK TIME/INJURED ON THE JOB

Abuse of sick leave: Never abuse sick leave. On the average, I would call in sick just one to two times a year. I viewed my sick time as an insurance policy; if needed, I could tap into it. If you get hurt at the fire department and can't go back to regular shift, the fire department will put you on light duty, consisting of Monday through Friday 0800-1700 hrs. This will be miserable for you. More than likely, you will be someone's slave by filing their paper work and running errands for them.

The insurance policy for me was just in case I got hurt off shift and did not want to do light duty or sustained a serious injury. I wanted to make sure I had enough time to cover it without dipping into my vacation time.

Some departments have a sick leave bank. This helps the seriously injured firefighter, who has exhausted his sick leave and vacation time, and now can use the sick leave bank. The hours in this bank are donated from other firefighters over the years and usually needs approval from a Sick Leave Bank Board. Here is the issue. If you abuse sick leave by keeping your sick time close to zero, then the Board will not likely approve your use of the sick leave bank. Don't be a sick leave abuser; keep a good amount in there just in case you ever need it. It is not a matter of if you will need it, but when.

Injured on the job and what to do next: Unfortunately, it is not a matter of if, but when you'll get hurt. Injuries are usually minor but can be major and or career ending. So, what do you do if you get hurt? Document, document, document! I cannot stress that enough. If you do not document your injury, then it did not happen. It doesn't matter if you think the injury is minor, or you don't want to be a wuss.

Everything must be documented for treatment to begin. Also, in case the injury worsens, there is a record of the occurrence. How do you do this? Just tell your supervisor. He or she will be put on notice that an injury has occurred and he will fill out the proper forms. This is required by law. You are covered by Workers' Compensation, if you need follow up treatment. If you find that your supervisor is too lazy to fill out the paperwork, then go over his or her head.

Do not screw around with this; it is way too important. Once the paperwork is filled out, get a copy of it and put it in a safe place. You will be surprised how often they come back and state, "We don't have any documentation of your injury."

Workers Comp has a way of treating the injured like a criminal. Do not tolerate this, especially if you feel you could have long term, or permanent effects. If workers comp turned down your claim, then you need to get a workers' compensation attorney. Ask your union who they recommend. In my area we have a few good ones whose entire practice is dedicated to firefighters and first responders. Report your injury, get the care that is needed and fight the fight, if you have to.

CHAPTER 24
CAREER TIME MANAGEMENT: SHORT AND LONG-TERM PLANS

Your first few years at your new career should be your busiest and most productive. This is where you need to have a shopping list of what you want and need to accomplish. The sooner you get these basic accomplishments done, the better and easier it will be for you. I have seen forty and fifty-year old firefighters attempt to go back to school to finish their college degrees or obtain enough college credits to take a promotional exam. This is very difficult to do when you are older. You probably will have family at home, or a side business you are trying to run. Be proactive early in your career and get the courses you need to be successful.

Listed below is a timeline to guide you in the beginning of your career. These are just approximations and can be adjusted accordingly.

Year One
- Make it through your probationary year

- Learn everything you can about your job

Year Two or Three

- Start taking your Special Ops Courses
- Begin Paramedic School

Year Three or Four
- Begin your Bachelor's Degree
- Start taking courses for future promotional exams

Years Four through Seven
- Study for promotional exams

This juncture is an important step in your career development. At this stage, you should be thinking about promoting. Some people don't want to promote. That's fine and should not be looked down upon. If you do want to move up in rank, you need to decide how far you want to go, what do you need to do to get there, and what road map you need to follow.

The following is the typical rank structure of a fire department:

Firefighter

Engineer

Lieutenant or Captain

District Chief or Battalion Chief

Chief of Operations or Deputy Chief (Usually Days)

Chief of the Department

What is important to remember is, the higher you go, the more political the promotion becomes, the less job security you have and the further away you are from the crews. Most people are happy with just achieving the rank of Lieutenant. This is because you are still on shift, are one of the guys, and still get invited to fire department functions. When you start getting up to Chief's rank, you are pretty much on your own and separated from the crew.

Chiefs are usually considered management and are programed not to be one of the guys. I never agreed with this because it has prevented some really good, smart people from moving up because they don't want to be separated from their buddies. On the flip side, I have known and worked for some incredible District Chiefs who always looked out for us and never sold their soul for advancement.

Your decision to seek promotion will probably change several times over the years, depending on priorities and what is going on in your life. If you remember anything I tell you, remember this: Do NOT promote to the Upper Chief's rank too quickly. This may be your ultimate goal, but don't do it too soon.

Here are five reasons, based on personal experience, why not to seek a Chiefs promotion too early in your career.

1. You are no longer one of the guys: Dinner conversations will be different; the crews will be standoffish, and your trips you take with your friends will be significantly reduced. In short, you will miss all the fun times with your friends.

2. Your job security will decrease: At a lower Chief's rank you will probably still be protected by the bargaining unit. Once you get to upper management, you are at the whim of the city leaders. This isn't a good or stable place to be.

3. You are more of an administrator then a firefighter: You will tend to do more paper work and manage personnel rather than getting your hands dirty and doing what you truly enjoy.

4. You won't like your job as much: Everyone I ever met who decided to promote too soon, absolutely hates their job.

5. Crews don't respect fast promoters: These upper positions command respect and very few can get it by moving up too fast.

Here is my recommendation for a sensible career and promotion time line. Keep in mind, this is purely subjective and should be based on what you and your family want. I know from working with numerous individuals over the years, this will work out well.

Year 1-6 Firefighter

Year 6-10 Work on promotion to Engineer

Year 10-20 Work on becoming a Lieutenant or Captain

Year 18- 20 Work on holding a Chief's rank

A very common strategy for many is to take the higher-ranking promotions to pad their pensions. This may or may not be the best strategy, but you and ultimately you, have to do what you feel is right for you and your family.

Getting ready for the promotional exams

Getting ready for a promotional exam is actually something you do the first day on the job. If you are a total idiot as a firefighter, then why would anybody want to work or respect you as they're commanding officer? You will be auditioning to lead your crew every day on the job. You really need to stay focused and remain professional at all times while mastering your job. You need to understand some people just won't like you. It is not because of something you do or say. You could be the nicest person in the world with a ton of street experience, and they still won't like you. Who cares! They are irrelevant. Do what you feel is right and treat others how you want to be treated.

Here is a quick career check off list that will add to your success

1. Do you have your degree and basic course work done years before the big exam?

2. Are you politically correct in front of the people that may promote you?

3. Have you followed the three-day rule?

4. Are you in shape and kept your fitness in peak form? Do you look, dress, and act the part long before you are promoted?

5. Have you worked on fire department committees? This will get you noticed and shows you are dedicated to your job.

6. Have you refrained from being a whiney complainer?

7. Have you stayed out of trouble and contributed to the team?

8. Were you a mentor to all the rookies that you worked with?

9. Have you taken a leadership role around the firehouse that everyone looks up to?

10. Have you helped others around you and always did what was right for your crew and not yourself?

These are only a few short recommendations for you. Someone else may come up with hundreds more. Most are common sense, but I think you get the picture.

CHAPTER 25
NOW THAT YOU'RE PROMOTED

I could write several books just on this aspect of the job. It covers a wide area from handling a problem employee to leading fire ground operations. This section will help you succeed in your job. Being an officer is not easy; it will be quite challenging, especially working with a bunch of type A personalities. You will encounter many egos with attitude, and some will inflict their own brand of wounds.

On the other hand, sometimes the officer is the one inflicting the wounds. Some are egotistical and narcissistic. The problem is, once the damage is done, it rarely can be repaired. So, from my vantage point, I am providing some suggestions to the new officer on how to avoid doing damage.

1. Help the crew with chores. You obviously don't need to do everything, but after a busy shift or big fire, the crew is tired. Don't lie in your bed while they are still outside in the hot sun or 3:00 a.m. working on clean up and preparations. Don't be afraid to get your hands dirty.

Being a young officer with Marion County Fire/ Rescue back in the day. Top left to right. Earnie Eubanks, Richard Broccolo, Lower left, Chief Wayne Futch and me

2. Tell your crew what you expect. If you tell everyone on your crew what you expect, then they will make sure your objectives are met. There will be no mystery about it. This could be anything from how you want the station to be cleaned to what tools you want taken off the engine during a fire. One Lieutenant I had, only cared about making sure there was extra toilet paper in his stall. So, every shift, I made sure spare toilet paper was always available. Easy fix.

3. Defend your crew when they are right. Get a spine! If you want to lose respect from your crew, then don't defend them when upper management is coming down hard on them with a new rule or accusation. Unfortunately, some officers are afraid of their own shadow or they are trying to get their next bugle and don't want to rock the boat. Defend them when they are right. Correct and lead them when they are wrong. Always, always put them first and not yourself!

4. Keep your emotions in check. DO NOT EVER scream, shout, point your finger or go toe-to-toe with someone. You must remain professional at all times. It's OK to be stern or to raise your voice occasionally, but keep your emotions toned down. You will earn the respect and cooperation of the crew.

5. Train your crew but don't go overboard. Get your company training done on weekdays before noon and not on weekends and holidays. I have seen new officers take their crews out several times a day, after dinner, and yes, sometimes at 2200 hours. This is totally ridiculous and makes the officer look like a self-serving idiot. Most crews love to train, just not at midnight. Respect their time and don't do training late in the evening.

6. Remain calm on calls. One of the best ways to garner respect from your crew and others listening on the radio is to remain calm on calls and speak in a controlled manner over the radio. This not only promotes calm in others, but it also lets them know you are in control of the situation. Others will respond favorably to this kind of calm leadership. If you are a screaming psycho idiot on the radio, then people aren't as likely to follow your lead.

7. Know your job. If you know your job and maintain a professional, calm demeanor, then the crews will follow and respect you. If you moved up too soon, or you aren't as knowledgeable as the task demands, or your crew needs, then you are in trouble. This is why I laid out a timeline on when I think it is appropriate to hold certain positions in the fire department.

8. Don't be afraid to ask for help from your crew. You are not going to know everything, and every situation is different. If you arrive on a scene, utilize the expertise of your crew to help mitigate what's going on. You are better off doing this than winging it and possibly making a mistake because your ego is too big to ask for help.

9. Don't show favoritism. Sometimes not showing favoritism is difficult. You will always have those one or two people you go to for advice or to accomplish a task. That is fine. What you want to avoid is disciplining someone for something but letting others get away with the same thing. Be consistent with everyone.

10. Use proper grammar and spell check. If you want to look like an idiot, then make sure you send your departmental email without using proper grammar or spell check. If you send an important message to the entire organization, you may want to have it read and checked by another person, or use one of the online services like "Fiverr." You can send them a document, and they will make the proper grammar corrections for you. They don't need to know that you had help, but they will know if you send the message without checking it first.

11. Make a decision. Don't be wishy washy or unsure of yourself. Some decisions will be difficult to make. That is expected in the life and death situations firemen encounter. But you are in a position to make decisions, even though some decisions are unpopular. There is nothing worse than having an officer in charge that is afraid of his own shadow. Even if your decision is wrong, at least you made one, and you won't look like you have a noodle for a spine.

12. Quickly correct the bad employee. There is nothing worse than having a bad employee, especially one with a bad attitude. This person is cancerous and will infect your entire crew. He or she needs to be isolated, shot down quickly, and set straight as soon as possible. Make him or her aware that you are the one in charge. Do not tolerate these people for a minute.

13. Let your crew eat first. You want to make sure there is enough food for them before you take yours. Never be the first in line or take the last piece of chicken at your second helping, especially when others are on their first. When dinner is over, help with the dishes. These firemen aren't your slaves. They are your crew; respect them, and they will, in turn, respect you.

14. Stay humble. Just because you wear the golden officer badge, does not mean everyone will look up and worship you instantly. It signifies that you should be honored to oversee a group of individuals who make your firehouse function as a well-run machine. Respect is earned not given. Just as easily as you will hold your crew accountable, they will hold you to the same or higher standard. Too often, especially in the higher ranks, officers think they are untouchable and act arrogant and superior to their crew members. This attitude is a bad idea.

The higher you go, the harder you fall. You are and will be held to a higher ethical standard. Over my twenty-five-year career, I have seen too many officers ruined by scandals. These scandals usually led to demotions, firings or criminal charges. Some of these people were good people, and I found it sad to see them end up this way. Some were great friends that got caught up with issues that they thought they were above.

One day when you leave the fire department and retire out after a long career. You want to leave with your head high, knowing that you always did the right thing for everyone that worked with you. You want to be that person, who after has retired, can show up to any firehouse and be welcomed back with open arms. You can't be a shit bag your entire career and then expect people too like you after you leave and want to hang out. In my personal book, it doesn't work like that, sorry. I judge you as a person based on how you treated people while you were in charge. That is a true test. To see if you had a spine when the shit hit the fan. Anybody can be a good person, while drinking a Pina Colada on a beach with an umbrella in it collecting a retirement check.

Remember, always do the right thing, and never put yourself first. The department is bigger and more important as an organization than you are as a person. The people that you think will be there to save your ass will be the ones to show you the door. Be humble when you seek promotion and enjoy the fruits of your labor.

CHAPTER 26
CAREER FINANCES AND RETIREMENT PLANNING

Retirement planning should begin on the first day on the job. Unfortunately, I have seen way too many firefighters live beyond their means, file for bankruptcy, and be forced to work several jobs just to make ends meet. When they are young and just out of high school or college, most people are usually broke and never really earned a real paycheck. Once they get hired, they are excited as money starts coming in. They plan for that dream home or truck they always wanted. Life is good. Finally, they think, they can now afford big ticket purchases and can't wait to start looking.

Another word of warning: Be careful. Slow down and think about what you are about to do. First, don't go out and get that $50,000 jacked up pick-up truck you always wanted or that condo on the beach. When making any financial decision, you must first ask, " Can I afford this purchase without working numerous jobs and getting overtime to pay for it?"

This is a major decision, as is any decision that will add debt to your portfolio. Too many times, guys make big purchases and thinking they can pay for them by working a few overtime shifts per month. Very bad idea. What if the overtime dries up? What if an emergency befalls the family? What if the department lays you off? I have seen countless times where there was more overtime than people to fill it. Shortly after that, overtime completely dried up. No overtime was available for anyone for several months. If you depend on overtime to pay your bills, you're in big trouble.

In the beginning of this book, I provided a realistic chart of expenses and income for a typical firefighter. The chart below varies in that, later in your career, you will have more income and want to purchase a more expensive house and vehicle. Interest rates vary but the chart gives a clearer picture of income needed to live a comfortable lifestyle.

$200,000 House@ 4.5% annual Interest

Mortgage	$1,013
Taxes	$ 250
Insurance	$ 125
Electricity	$ 250
Phone/Cable/ Internet	$ 150
Repairs	$ 100
Miscellaneous	$ 200
Vehicle	*$30,000*
Payment	$ 450

Gas	$ 250
Insurance	$ 100
Food	$ 600
Fun	$ 350

Total take-home needed per month	$ 3,838
Total take-home needed per year	$46,056 (Does not include savings)

As you can see, just to purchase a modest $200,000 home, you will need a monthly net of $3,838. When you are new to the job, this is difficult to get. You only have a few options such as buy a cheaper home (be sure it's in a good neighborhood), cut back on monthly expenses (buy a cheaper car), or rent an apartment.

Some people get married so they can split expenses. This may work but we didn't even mention starting a family. So, what are you to do? The following suggestions worked for me and they might work for you. They aren't easy; they require discipline and restraint when it comes to setting a budget and sticking with it. Here are some important considerations.

1. Should I rent an apartment or buy a home? My advice is this. If you are new in the town that hired you, I would rent for your first year for these reasons. First, you have absolutely no idea which are the good schools. Second, you don't know which neighborhoods have the best home values and which are in transition and improving. Finally, you will want to live in areas that offer the best amenities for that city. Besides, you may not like your particular fire department and want to apply to another.

You really want to take your time with this and do your homework. I can't stress that enough. If you are not sure where to rent or buy a home initially, then stop by the local fire station that covers the apartment complex or neighborhood that you are thinking about buying in. They can tell you exactly where the high crime calls originate and the areas to avoid. During the day, an apartment complex may seem really nice on the surface. That's because the animals are sleeping during the day because they were out causing problems at night. I'm not kidding. Ask any firefighter who works in any area. They will tell you exactly what you need to know.

Once you start working, you will easily and quickly identify the good and bad areas. You will also float in and out of several stations around the city and determine first hand where the good area are located. So basically, if you're new to the area, take your time to buy a home.

2. Will my fire department pension be enough? After retirement, if you are counting on the fire department pension to give you a safe monthly income that will keep up with inflation, then you are sadly mistaken. Most pension systems do not have cost of living increases.

So, what do you do to supplement your income? From the first day on the job, you need to have your own retirement plan, just in case something goes wrong with the state or your city's pension system. This won't happen, you say? When the economy is bad, or city officials poorly manage the employee pension fund, one of the first things they cut is the retirees' pension.

This happened to Detroit right before they declared bankruptcy in 2013. The firefighters who were retiring, along with those already retired, took major pension cuts. Think about that for a second. You're used to having a certain amount of money coming in every month to pay bills. All of a sudden, you get a notice that your pension just decreased 30- 50 percent, and there is nothing you can do about it. Because of this, you need your own pension plan. This will be an insurance policy in case of a disaster or to just boost your yearly income in retirement. There are a couple of practical ways you can do this.

3. Is Deferred Compensation, the power of compounding interest, right for me? A 457 Plan is actually a wonderful way to build a huge nest egg over a long period of time. The key to this is the power of compounding interest every year and putting in a regular amount of money into the fund. The best part about it is, it is done with pre-tax money. So basically, you are putting tax free money away into a compounding savings account. You paid no taxes before putting the money in, and you pay no taxes while you are earning money. Your tax liability will be a lot lower in your paycheck because it's all tax free until you start collecting it.

As of this writing, the yearly government contribution is $18,500. So basically, you can save up to $18,500 per year, tax free. Listed below is an example of the power of compounding interest.

Monthly Savings. $250 or $3000 per year times 25 years= $75,000

Average Yearly Interest 7%. Your actual contribution was $75,000. With this formula, you will have earned $190,000 with the compounded interest. The best part is that the money is tax free until you withdraw.

If you can't save $250 month, then do anything you can, even $50 per month and increase it from there. Some people take their yearly raise or their overtime and put it in. I strongly urge you to do this. Don't worry about the day-to-day fluctuation of the Stock Market. History tells us over the long term, you will be fine.

4. Should I purchase a rental income property to supplement my income? Maybe you never considered this option, but I want to recommend purchasing a rental income property. If you choose this route, you MUST know what you are doing. Over the years, I have owned a long-term rental home and a vacation rental. Both generated income streams, gave me excellent tax benefits and appreciated over the years. The key considerations for these properties and being a successful property owner is fivefold: purchase price, high occupancy, affordable upkeep, desirable property location, long term appreciation potential.

A. Purchase these below market value, preferably a fore-closure or some type of distressed sale. Make sure you have equity in it immediately after taking ownership. This is important in case you have to off load it in a short turn-around time.

B. If the rental does not rent out or there is a long period of vacancy, make sure you can afford the payments.

C. Make sure you can afford the up-keep and all mainte-nance issues that may come up.

D. Purchase in a desirable location. This will get you maxi-mum appreciation and maximum rental income. Buying a property in a rundown area may look like a bargain on the surface, but you will have little to no appreciation, maybe even have a depreciating asset. Plus, you will have nothing but problems collecting rent and fixing a disaster after the tenants move out. Cheaper is not always best.

E. Time. There are no get rich quick strategies worth risking your money on. You must give your investment properties time to mature in value, especially if you purchased them correctly. Here are the benefits of holding these properties for the long term.

You bought a home for $150,000, rented it out for several years, and paid if off. You now decide to sell it years later for $225,000 and pocket the entire amount.

Your second vacation rental cost you $175,000. You also rented it out over numerous years and paid off that mortgage, also. You now decide to sell that asset which is now worth $276,000 and pocketed the sale price. Not bad right? So, here's how you fund your retirement: Check this out.

Rental Home Sold.	**$225,000**
Vacation Rental Sold.	**$276,000**
Deferred Comp.	**$190,000**
Total.	**$691,000**

That's $691,000 in real estate assets sold and your Deferred Comp Savings Plan, plus your yearly pension. You may have to pay the tax man, but you have earned enough to cover this expense. Your savings still look pretty.

The above scenario is not impossible. The key to this success is to start early in your career, don't live above your means, and be financially disciplined. You don't need that $400,000 home when a $250,000 home will suffice. You don't need a new

$50,000 pick-up when a used vehicle with 30,000 miles is just as good.

Don't go crazy, make good solid financial decisions with your spouse and good things will happen.

How do you know it's time to retire? You read this book, and you had a wonderful thirty-year career. You are getting older, smarter, and just want to enjoy life for your remaining years. When is the time right to say goodbye to your career and live the retired life? Here are five clues to help you decide if retirement is a good idea.

1) You have a bad attitude. Throughout your career, things were great. You were happy, laughing, and looked forward to coming on shift to hang out with your buddies. As time went on, you noticed the job wasn't fun anymore and things started to change. You became less tolerant. If you have your time in and notice these signs, then it's time to retire.

2) Your home is paid off. Please don't go out and buy a new home with a $300,000 mortgage just before you retire. If you do, you may have to work another thirty years.

3) Your kids are out of school and moved out. This will be the biggest raise you get. No more child expenses!

4) Most of your big bills are paid off.

5) You have enough extra money in savings and supplemental retirement accounts so you can travel, if you want to.

If you aren't smart with your money or your life, you may be stuck having to keep working. But if the work you do does not excite you, and you don't look forward to being at the station, then it may be time to leave.

CHAPTER 27
YOUR LEGACY

You are now near the end of your career. What will your legacy be? How will the department remember you? To me, this was extremely important. Some people don't care how they will be remembered or what they leave behind. I wanted people to say just two things about me around the dinner table after I left: "He was a good guy," and "He was a good fireman." That's all I cared about because I couldn't control anything else.

I wanted to make sure after I retired that I could walk into any firehouse in Orlando and the guys would greet me as if I never left. I wanted to leave with my head up. I wanted my reputation to be free from the controversy or scandal that so many others leave behind. I think I succeeded. To this day, I still get calls from the crew, invitations to fire department functions, and get asked to join committees. For me, there is no better feeling than to be remembered and valued by your co-workers.

Me, after getting my butt kicked at a fire

Your legacy is up to you. Treat your peers with respect, act courteously toward your co-workers and supervisors, and present yourself as a professional. Only you can control this. Stay true to yourself, and you, too, will leave a great legacy.

CONCLUSION

This book has given you everything you need to "Survive the Firehouse." I shared the knowledge gained from a long firefighter career. I made mistakes or watched others make theirs. You are now better prepared to handle the challenges before you. As you begin and end your journey as a firefighter, you need to understand this: You won't be perfect; you will make mistakes, and not everyone will like you. You will have bumps and forks in the road to navigate around. If you keep true to yourself, always do what is right for your crew, your department, and yourself, then I promise the legacy and impact you will leave behind will be a positive one and one that others will strive to emulate.

Good luck to you on your journey.

QUICK REFERENCE GUIDE

Orientation

- Arrive 1 hour early
- Be the first to do everything, cleanup, hose loading
- Always be up front when running
- If working out or after an evolution, never let them see you tired, keep head up, don't show weakness
- Don't be arrogant about your prior experience
- Keep your mouth shut, talk less, listen more
- Don't tolerate smart asses in your class. Make sure you rip them back or just sit there and smile. Never show weakness
- Don't tell anyone who you know. They don't care; they will find out on their own.
- Down play everything
- Never ever say in orientation or on shift, "You want to be on the Tower or Heavy Rescue"

Rookie at the firehouse

- Be at station at 6:30 am
- Put up and take down flag
- Make coffee
- Make tea in morning and before dinner
- Replace ice in morning and before dinner
- Be the last to go to bed...Be the first to get up
- Tighten up kitchen before you leave shift and before you go to bed
- Be the first to get up from the table and do dishes
- Be first to answer phone
- Be first to put aluminum foil on dinner plates if another unit gets a call
- Always help in the kitchen and other areas of station, even if not assigned
- Be the last to finish station cleanup
- Wash engine even if does not need it
- Help other units clean their truck, but always do yours first
- When crew is inside station, make sure you are outside learning about your truck and where everything is
- Always have your SOP's and other FD books open on table
- Don't sit in recliner during day or at night with crew, stay in kitchen reading your SOP's or at a desk
- Always help in kitchen
- Always help cook dinner
- Know where your equipment is stored

Bathrooms

- Throw away old soap, replace with new and leave in package
- Have extra toilet paper in stalls; keep in rapper
- Leave plenty of paper towels
- Pour Pine Sol in toilets and sinks; if none available, use bleach. It needs to smell clean.
- Mop bathrooms with bleach (Not too much) and water

Basic simple recipes for those challenged in the kitchen

I have included some tasty recipes for the beginner firehouse cook or the lazy guy that wants everyone to think he really can't cook. Most of these were popular at my firehouses. You will notice that most of the ingredients don't have to be precise. Just do an approximation; this isn't rocket science. As you cook some of these staples, feel free to put your own spin on the recipe. Next time you prepare the dish, just add or delete what didn't work. The crew won't know the difference, and you'll be a kitchen hero.

BBQ Chicken.

There is usually somebody on a diet, wants to eat healthy or whatever. You can never go wrong with BBQ Chicken.

Take some chicken breast, split it down the middle and marinate them in Italian dressing for a few hours. Season with whatever you like. Grill on median heat about 30 minutes or until it's cooked thru. Brush BBQ sauce on the chicken only the last 5 minutes. Serve with a good salad, baked potato and garlic bread that you can buy frozen.

Station 10 Chicken tacos.

The Tower Truck at Station 10 used to cook all the time, I didn't mind it since I really enjoyed cooking, but the other guys used to bitch all the time about what we cooked, how we cooked etc. Well, what did we do to correct that? We cooked Chicken Tacos for something like 10 shifts straight, until they quit bitching.

Take some chicken breast, split it down the middle and marinate them in Italian dressing for a few hours. Grill on median heat about 30 minutes or until it's cooked thru. Slice chicken in small strips. Serve with flour tortilla, lettuce, tomato, sour cream, refried beans, black beans and yellow rice.

Crock Pot Chicken.

This is always good, especially if you won't have much time cooking that shift because of training etc. Crock pot cooking is also perfect for a single company house or very small crew. You can't screw it up plus it cooks very slowly over several hours. Give it a shot.

Take chicken breast or thighs and put in a crock pot. Add a large can of cream of mushroom soup, a package of Lipton Onion Soup mix and 1 whole onion. Takes about 3Hrs. Serve over your favorite rice and Mauro's Pasta Salad.

Mauro's Pasta Salad

This is a very simple side dish that is also filling. Doesn't really take long to put together and it is very flavorful. Put it together in the morning and let it sit in the fridge all day. This really condenses the flavor.

Boil 2 lbs. of pasta. Rotelle or Ribbon pasta works best. Drain and add a whole bottle of Newman's Caesar salad dressing. (Not the creamy kind) Add small chopped carrots, celery, onion and about 1/2cup of grated Parmesan cheese. Stir and place in fridge all day stirring occasionally. Just before serving, add another 1/2 bottle of dressing and sprinkle a little more Parmesan cheese on top

Chicken Curry

My buddy, Lieutenant Dave Calder of Marion County Fire/Rescue used to cook this dish all the time. It was tasty, easy to put together and cheap to make.

Take 4 chicken breast and place in a pan. Add 1 cup of mayo, 1 cup of shredded cheddar cheese, a can of cream of mushroom soup and 1 lime. Add 5tbsp of powdered curry. Cook covered in oven at 375* for 1 hour. Serve over Jasmine rice with a side of Cesar Salad.

London Broil

Marinate in Teriyaki for several hours. Grill on medium/high heat until inside is medium rare. Serve with Matt Holcomb's World- Famous Garlic Bread (recipe below) and Cowboy Beans. Slice London Broil very thin. Add a salad.

Chicken Pot Pie

This is a great dish for a cold winter day or any time. Just make 2 pies for 4 people or an entire large oven pan for 6 people.

Brown cubed chicken breast add salt and pepper. Once brown, place in bowl with cream of chicken soup, frozen bag of vegetables and small cubed cooked potatoes. Place in frozen pie crust. cook at 375* for 45 minutes. Last 15 minutes, put biscuits on top of pie and cook until brown.

Cowboy Beans

Brown a small package of Ground Beef and 2 packages of Jimmy Dean Sausage, Sage or Maple flavored. Add large chunks of 1 large onion. Add 4 cans of Baked Beans, and about a cup of Maple Syrup. Simmer on low for about an hour.

Meat Loaf

There are literally hundreds of variations to a homemade meat loaf. However, you decide to put it together, just make enough of it for leftovers for lunch next shift.

Buy 1 large package of ground beef. Place into a large bowl, add 4 eggs, 1 package of Lipton onion soup mix and 1 cup of ketchup. If it's too soupy, add 1 handful of bread crumbs. Mix well and place on a flat pan. Mold it into an oval looking football. Add a bunch of ketchup on top of Meatloaf. Cover with aluminum foil, place in oven at 375* for 1 hour. Serve with mashed potatoes, green beans and corn bread. Brown 1lb of bacon first with 1onion and simmer with green beans from the can.

Engineer BJ Shanks Enchiladas

BJ is a City of Orlando Firefighter and used to make the best Enchiladas. They were very filling, cheap to make and there is almost always left overs. Get some help filling the Tortillas and make a lot of them for the crew.

Enchiladas for 5 firefighters

ingredients:

- 3 lbs ground beef

- 10 or 15 burrito size flour tortilla

- 16 oz Mexican cheese

- 48 oz red enchilada sauce

- head of lettuce

- tomato

- onion

- sour cream

- cilantro

Preheat oven to 350°

1. brown ground beef and drain

2. add 24 oz of enchilada sauce to cooked meat and let simmer 10 min

3. add 8 oz cheese to meat mix and stir to melt

4. pour couple ounces of sauce into bottom of 9 x 13 casserole pan

5. scoop spoonful of mixture into tortilla shell and roll into burrito size placing into casserole pan (do multiple enchiladas till pan is full about 10)

6. cover all enchiladas with remaining sauce and top with remaining cheese.

7. place in oven 30 minutes

8. garnish with cilantro and preferred vegetable toppings, sour cream

Serve with rice and refried beans

Rice secret. Never open the lid and make it tight, even if it means using tinfoil

Mauro's Bruschetta Bread

Talk about a flavorful little bread dish. Just from a whim, I put this together one day with some of my favorite ingredients. I really was not sure how it would turn out. The crew loved it and was in hot demand whenever I got floated out to different stations. One guy told me it was the worst garbage he ever ate, but he went back for 5 helpings. Go figure.

Dice a few tomatoes, a handful of fresh basil and a handful of red onion. Add salt, pepper and olive oil. Stir regularly and place in fridge all day.

Take a loaf of French or Italian bread, cut long ways down the middle. Spread melted butter and sprinkle a little garlic powder on it. Take shredded mozzarella and spread on bread and bake until crisp. Take out of oven and cut to serving portions. Take your tomato topping and spoon on top of bread. Drizzle a little Balsamic Vinegar on top, if you like.

Engineer Matt Holcomb's World-famous Garlic Bread recipe

Engineer Matt Holcomb of the City of Orlando Fire Department used to make this incredible Garlic Bread. It was sure to get your cholesterol up and harden your arteries. But it was incredibly tasty and you can make it with anything.

- 1 loaf French bread

In a medium sauce pan, bring to a boil and melt the ingredients together.

- 1- 1/2 sticks salted butter

- 1 table spoon minced garlic

- 1/4 cup Italian seasoning

- 1/4 cup basil seasoning

- 1/4 cup oregano

- 1 cup olive oil

- Shredded cheese

Optional: Slice tomato in thin slices and lay on top of bread.

Cut bread into halves

Spoon melted ingredients onto the halves and then tomato's and cheese

Bake at 350* until the bread is nice and crisp

Slice into pieces and enjoy

Frank Bello's Arthur Avenue Chicken Cutlet Parmesan

Frankie is a retired, highly decorated FDNY Firefighter who worked at one of the busiest Engine Companies in New York City, 88 Engine in "The Bronx."

Frankie was known throughout the city for his famous Arthur Avenue Chicken Cutlet Parmesan. What made this incredible dish so awesome was the fresh ingredients he used from the world-famous Little Italy district in The Bronx called Arthur Avenue.

Start with a heavy gauge gravy pot, chop 5 gloves of garlic and 1onion coarsely, the amount would depend on how much sauce you're making, assume you are using 5 cans. Put about a quarter inch of olive oil in pot, brown garlic, and onions. (do not burn)

Frankie uses "Tuttorosso" canned whole tomatoes. You can use another brand if not available. Put tomatoes in blender, blend on low. Be careful not to break the seeds. Add to pot.

Add some sweet Italian sausage, pork, fresh meatballs and add to the pot. Cook for at least 3 hours, very low, stirring often.

Chicken Cutlets

Take chicken cutlets, do not filet, if they're very thick you can pound them down a little bit, batter chicken in egg and bread crumbs. Fry chicken till brown. Put on paper towels to soak up excess oil.

If you can find it, Frankie likes to use fresh mozzarella cheese. Place cheese in the freezer for a short period to firm it up before you grate it.

Sprinkle cheese on top of Chicken Cutlets, put in oven until cheese is melted.

Serve with Angel Hair Pasta. After pasta is cooked, pour sauce over pasta and cutlets

Serve and enjoy!

ABOUT THE AUTHOR

Mauro Porcelli is a retired City of Orlando Firefighter/ Paramedic with twenty-five years of fire service experience.

He started his career with Marion County Fire/Rescue in 1988 where he advanced quickly through the ranks of Lieutenant and Captain. At twenty-three, Mauro was one of the youngest, highest ranking officer for a professional firefighter in the State of Florida, holding the rank of District Commander. He eventually left Marion County for the City of Orlando Fire Department from which he retired.

Mauro is a highly decorated firefighter, receiving several commendations, including one from The United States Congress, Florida House of Representatives and Florida State Governor's office, and two years as Firefighter of the Year with commendations from the Veterans of Foreign Wars. Mauro also received two medals of Distinction for Valor and Heroism in the line of duty.

Mauro worked in some of the busiest stations and roughest neighborhoods in the state. His life long experience and "street cred" earned him the respect of his peers and professionals alike.

During his career, he gave selflessly of his time to mentor rookies in the firehouse as well as others seeking his advice on how to succeed and advance in his beloved profession. His book grew out of this desire to help others as they "Survive the Firehouse" This book is a collection of his insights, foibles, pet peeves, overall wisdom and tips on how to make it as a firefighter.